Gender and Educational Philanthropy

List of Previous Publications:

Alice E. Ginsberg, Joan Poliner Shapiro, and Shirley Brown, *Gender and Urban Education: Strategies for Student Achievement* (New York: Heinemen, 2004).

Gasman, Marybeth and Katherine Sedgwick (Eds.), *Uplifting a People. Essays on African American Philanthropy and Education* (New York: Peter Lang, 2005), 204 pp. **Winner of the Association for Fundraising Professional's Skytone Ryan Research Prize.

Gilpin, Patrick and Marybeth Gasman, *Charles S. Johnson: Leadership behind the Veil in the Age of Jim Crow* (New York: State University of New York Press, 2003), 318 pp. Foreword by David Levering Lewis, author of *W. E. B. Du Bois: The Fight for Equality and the American Century*.

Gasman, Marybeth and Sibby Anderson-Thompkins, *Fund-Raising from Black College Alumni: Successful Strategies for Supporting Alma Mater* (Washington, DC: CASE Publications, 2003), 119 pp. **Winner of the H.S. Warwick Award for Outstanding Published Scholarship on Philanthropy.

Forthcoming 2007:

Gasman, Marybeth, *Envisioning Black Colleges: A History of the United Negro College Fund* (Baltimore: Johns Hopkins University Press, 2007), 290 pp. Foreword by John Thelin, author of *A History of American Higher Education*.

Gender and Educational Philanthropy

New Perspectives on Funding, Collaboration, and Assessment

Edited by

Alice E. Ginsberg
and
Marybeth Gasman

GENDER AND EDUCATIONAL PHILANTHROPY
© Alice E. Ginsberg and Marybeth Gasman, 2007.

Softcover reprint of the hardcover 1st edition 2007 978-1-4039-7533-1

All rights reserved. No part of this book may be used or reproduced in any manner whatsoever without written permission except in the case of brief quotations embodied in critical articles or reviews.

First published in 2007 by
PALGRAVE MACMILLAN™
175 Fifth Avenue, New York, N.Y. 10010 and
Houndmills, Basingstoke, Hampshire, England RG21 6XS
Companies and representatives throughout the world.

PALGRAVE MACMILLAN is the global academic imprint of the Palgrave Macmillan division of St. Martin's Press, LLC and of Palgrave Macmillan Ltd. Macmillan® is a registered trademark in the United States, United Kingdom and other countries. Palgrave is a registered trademark in the European Union and other countries.

ISBN 978-1-349-53599-6   ISBN 978-0-230-60308-0 (eBook)
DOI 10.1057/9780230603080

Library of Congress Cataloging-in-Publication Data

Gender and educational philanthropy : new perspectives on funding, collaboration, and assessment / edited by Alice E. Ginsberg and Marybeth Gasman.
    p. cm.
Includes bibliographical references and index.

    1. Endowments—United States. 2. Sexism in education—United States. I. Ginsberg, Alice E. II. Gasman, Marybeth.

LC243.A1G37 2007
378.1'06—dc22   2006049649

A catalogue record for this book is available from the British Library.

Design by Newgen Imaging Systems (P) Ltd., Chennai, India.

First edition: April 2007
10 9 8 7 6 5 4 3 2 1

Contents

Foreword
Andrea Walton — vii

Acknowledgments — xv

Introduction
Alice Ginsberg and Marybeth Gasman — 1

Part I Definitions of Gender Equity

One   Talking about Gender Equity and Education:
      A FrameWorks Message Memo — 17
      Joseph Grady and Axel Aubrun

Two   Gender Matters: Funding Effective Programs
      for Women and Girls — 33
      Molly Mead

Three Grantmaking with a Gender Lens — 67
      GrantCraft

Four Bringing Funders Together to Talk about
      Girls: A Roundtable Discussion — 89
      The Valentine Foundation

Five Saving Black Boys — 109
      Rosa Smith

Six   Including Boys in Our Conversation about
      Gender and Justice — 117
      Michael Reichert

Seven Supporting Boys' Resilience: A Dialogue with
      Researchers, Practitioners, and the Media — 133
      Allyson Pimentel

Part II Collaborations and Program Assessment

Eight Power and Possibilities: Collaborative Fund for
       Youth-Led Social Change — 157
       *Ami Nagle, Marisha Wignaraja, P. Catlin Fullwood,
       and Margaret Hempel*

Nine	Young Women for Change *The Michigan Women's Foundation*	177
Ten	Collaborations for Gender Equity in the Context of Policy and System-Wide Change: An Interview with the Editors *Shirley Mark, Alice Ginsberg, and Marybeth Gasman*	187
Eleven	Sisters Empowering Sisters and a Case Study: Girl World *The Girl's Best Friend Foundation*	191
Twelve	Gender Equity in Urban Education: New Relationships between Funding and Evaluation *Alice Ginsberg*	203

Notes on Contributors	235
Supplementary Bibliography	241
Index	255

Foreword

Andrea Walton

On March 20, 2005, the Women's College Coalition (WCC), a nonprofit devoted to "the education and advancement of women," sponsored a full page advertisement in the *New York Times*.[1] Underneath the bold attention-grabbing headline "**Women Can't Do Science?**" two black and white photographs formed part of the advertisement. One picture, dating from the late nineteenth or early twentieth century, showed a group of young women (their floor-length, high-collared dresses reflecting the era's sartorial customs) working together in a science laboratory. The other picture, depicting a contemporary scene, showed two female students, wearing white lab coats and safety goggles, performing an experiment with a piece of modern scientific equipment. Marking the observance of women's history month and appearing while the controversy on gender differences and scientific ability sparked by Harvard's president Lawrence Summers was still ablaze, this high-profile advertisement, endorsed by the WCC's membership, sought to help refute the ill-founded but common assertion that women *can't do science* by marshaling evidence drawn from history—in this case, photographic images and a brief note about the record of female college graduates in science—suggesting that women have, in fact, always *done science*.

The WCC-sponsored ad—both in its ideas and its strategic timing vis-à-vis the Summers controversy—reminds us that despite the massive advancement of women in most fields during recent decades and the ample evidence documenting women's intellectual ability during the past century and a half, old arguments questioning women's intellectual ability still linger. Indeed, the WCC's efforts to fight back on women's behalf, by confronting current efforts to revive old nature versus nurture arguments, is compelling evidence of the salience of gender dynamics in education and the vigor of gender debates today. But beyond this, the WCC's action provides an example of the vitality of philanthropic action in U.S. society and the role philanthropy can play in helping to alert the public about important issues and in both testing new ideas and in challenging barriers to equity. Of course, while private giving to education in the United States cannot be viewed uncritically and certainly has not been without its pitfalls and limitations, it must be noted that educational philanthropy—the efforts of individual donors, voluntary organizations, nonprofits, and, perhaps most visibly, foundations—has been an instrumental tool, both in the distant past and in recent decades, in the effort to broaden access and to promote gender equity. Moreover, philanthropy has been an avenue by which not only women but also other excluded or marginalized groups have taken control of their own education.

To my mind, the WCC ad, albeit providing but a glimpse into one particular nonprofit group's larger agenda devoted to women's education, points to an important aspect of the problem of gender in the history of U.S. education: that the problem has persisted and that the discourse about gender is ever-changing. More to the point, in addition to underscoring the major issue of women's visibility in science, these photographs—simply by their juxtaposition—remind us that contemporary issues and concerns are deeply rooted in history and that history may provide a context for understanding the present. Therefore, bearing in mind the importance of context for bringing meaning to the "nuts and bolts" details of contemporary projects and initiatives on gender in education, I focus my observations on two fronts: first, on the long-standing ties between educational philanthropy and gender issues (the early period when women were fighting for access); and second, on a few prominent trends that provide the social and intellectual backdrop for the philanthropic initiatives examined in this volume (the recent decades when a new set of issues arose once women became an integral part of educational institutions).

Observations about Early Trends

At the outset, our early schools and colleges, primarily founded and bolstered by philanthropy, excluded females. In response, women and their supporters had to fight for women's rights to education and, accordingly, the tradition of educational philanthropy was one of the avenues for pursuing their goal. Using the power of private funds, volunteered time, and collective action to broaden ideas about educational access, to build new institutions and change old ones, women strove to secure a meaningful education for themselves and for girls. While by the 1790s the need for female schooling had begun to gain hold, by the 1830s and 1840s girls attended common schools with boys and reform-minded women like Mt. Holyoke's Mary Lyons had successfully established institutions to provide young women with the type of liberal arts study that was readily available to men. Other prominent examples of women's educational philanthropy and institution building to improve educational opportunities for women followed. Back then, as now, the approaches varied. Some efforts embraced gender separation. For example, Sophia Smith's generous bequest led to the founding of a women's college named after her in 1875. Women helped support the founding of women's coordinates at elite men's colleges—for example, Radcliffe at Harvard in 1879 and Barnard at Columbia in 1889. In response to racial segregation, women built institutions devoted to educating African American women. Mary Bethune's Daytona Normal and Industrial Institute for Negro Girls, founded in 1904, is a case in point. Other efforts by women sought to forge inroads into male-dominated institutions by introducing coeducation. One thinks of Mary Garrett's gift to open the medical school at Johns Hopkins to women in 1893. Yet other efforts, like those of Berkeley regent Phoebe Hearst in the late 1890s and early 1900s or of organizations such as the American Association of University Women (AAUW), sought to change the culture of coeducation to be more supportive of women (Solomon, 1985; Rosenberg, 1982).

But women and their supporters, much like advocates of gender equity today, often encountered vocal resistance from critics who saw a push for gender-equitable policies and practices as an assault on standards or an effort to politicize (or feminize) the school or college campuses. Indeed, as women's collegiate participation began to rise a spirited "backlash" ensued—to which women responded. In 1885, the fledgling AAUW published a study of women's health statistics in order to refute the ideas of Harvard physician Dr. Edward Clarke's, whose bestselling *Sex in Education* (1873) had popularized the argument that the demands of collegiate-level studies posed a danger to the female reproductive organs. (One cannot miss the irony that this same elite institution is today at the heart of gender debates.) By the early decades of the twentieth century, women's outstanding scholastic record had instigated another moment of backlash that women had to negotiate. Perhaps most notable was the University of Chicago's attempt to "retreat from full coeducation," by arguing for the segregation of undergraduate classrooms by gender (Solomon, 1985, 58).

While women were advancing into higher education, girls were also making great strides in attending public schools and normal schools and afterward entering the teaching profession. And by the early 1900s, the record of girls' achievement in schooling and the predominance of women in teaching had prompted a backlash no less vigorous than was the case in higher education. Critics warned that the feminization of high schools created an inhospitable environment for the male students and helped explain their early departure from school. The crisis was dubbed the "boy problem"—a phrase that will resonate strongly for readers of this volume as it explores recent debates concerning gender and educational philanthropy (Tyack and Hansot, 1992).

Poignantly, in the first decade of the twenty-first century, we once again find ourselves at a decisive moment. The history of women and girls in the classrooms has changed how we think about gender, as have a number of philanthropically funded initiatives, especially foundation-funded projects and programs, related to gender in education—notably, for instance, a push for a more inclusive curriculum and the founding of research centers devoted to the study of gender. But some observers today, as indeed did others of earlier generations, believe that many of the initiatives to aid girls and women in the name of equity have in fact launched a "war against boys" and men and given rise to a new type of gender gap (Sommer, 2000). By contrast, others argue for a broader concept of gender and regard attention to the special needs of boys as part of a continuum of our efforts to democratize education and broaden access, not as a trade-off or a zero-sum game but as a win-win situation for both boys and girls, men and women. In this context, funders face a particular challenge and are keenly aware that their decisions will, inevitably, shape not only opportunities for students but also the opportunities that students envision for themselves.

Observations on Recent Trends

The timeliness of this volume and the context for understanding the programs and initiatives discussed in this anthology may best be appreciated against the backdrop of crucial developments and shifts in thinking about the three subjects of this book—gender, philanthropy, and education—during the past two decades or so.

With respect to gender, the policy discussions have widened from an initial concern with challenging the discrimination that women faced to an awareness of the gender expectations confronting all individuals and, consequently, a concern with equalizing and improving opportunities for both sexes throughout life. On another related front, thinking about the sexes has acquired a new relevancy in education as disciplinary theorizing has been increasingly attentive to issues of race, class, religion, and ethnicity and to the diversity among girls and women, boys and men. Further, gender analyses have compelled us to examine tacit assumptions undergirding the status quo, helped us reconceptualize old paradigms, and brought new insights into the study of education, history, economics, international development, and, indeed, into nearly every field and profession. Finally, building upon the foundational work in women's studies, a new wave of scholarship has pointed to the constructed nature of masculinity, to the challenging cultural expectations and pressures facing boys and men, and, most recently, to the fluidity of gender identity and to an exploration of the perspectives and experiences of transgendered individuals.

Within the field of philanthropy some of the most interesting developments and those most relevant here have centered on foundations and their activities. Having incurred their share of criticism for elitism, foundations, to varying degrees, have begun recently to strive for greater transparency in their own actions, to encourage a more participatory culture of giving in society, and to build a knowledge base aimed at making philanthropy more effective and more responsive to social needs. For example, since the early 1980s, a number of foundations have supported efforts to learn more about and to nurture the philanthropic traditions of various groups—notably among them, women and communities of color—and, further, foundations have funded university-based centers and professional organizations devoted to improving philanthropy. Moreover, as foundations have proliferated in recent decades, the types of foundations and their activities have also become more varied. One can point, for example, to a rise in family foundations, community foundations, and more radical social change foundations—many of which have more diverse governing boards than has traditionally been the case in large national foundations and have shown interest in issues related to gender in education.

In the broadest of terms, the field of philanthropy has experienced changes that have increased its own awareness of the cultural context in which it operates. Mindful that philanthropy is intimately connected to issues of power and empowerment, some philanthropists have tried to adopt a less hierarchical relationship between donors and recipients and (partly reflecting the influences of feminist and grassroots movements) embraced a framework of working *with* instead of working *for* recipients (to invoke a crucial distinction emphasized by noted educator-philanthropist Grace Dodge). We have moved from talking mainly about women as recipients of philanthropy to an appreciation of how women's leadership as philanthropists—visible in the efflorescence of women's giving circles, women-led community funds, women's foundations, and many of the initiatives detailed in this volume—can make inroads in both gender-specific and more general areas of giving, such as responding to the urgent national concern with better educating younger generations about philanthropy and civic engagement.

While at present only a small fraction of funding directed each year to education goes directly to projects for women and girls, the energy, resources, and creativity being directed to philanthropic initiatives by and for women today has the possibility of invigorating the pursuit of gender equity in education—and of benefiting the culture of U.S. philanthropy more generally.[2] Indeed, the growing power of the women's philanthropy movement and the more democratic, reflective, and collaborative trends associated with this movement are not insignificant given the predicted impact of an imminent transfer of wealth on women's financial means and on the future potential of educational philanthropy (Havens and Schervish, 2003).

Finally, it is important to note that this edited volume on gender and educational philanthropy appears at a pivotal moment in our national conversations about school reform and about equity and excellence in education. Reflecting a number of education reforms that were sparked by the publication of *A Nation at Risk* in 1983 and fueled more recently by the controversial No Child Left Behind Act of 2001, schooling has assumed an ever greater prominence on the national agenda and received more intense scrutiny and interest from the public, media, government, business, and philanthropy. This additional attention is warranted. For, although as a society we have historically invested an abiding hope in our schools (one recalls that as early as the 1840s Horace Mann touted common schools as the "great equalizer"), there is still an untenable gap between our ideals and the true measure of equality in our classrooms. Indeed, a substantial body of research suggests that schools—where our nation's children and youth spend considerable hours—often reflect and in certain ways help to perpetuate (whether overtly or subtly) a great many social cleavages and inequalities, including gender biases.

In response to such troubling research findings about educational outcomes and to new political demands, some national funders have begun to shift their funding priorities and interests from higher education to schooling. This new emphasis both reflects and has deepened our ever increasing awareness of the crucial experiences of a child's formative years, the benefits of early intervention, and the ability of philanthropy—an external force that can transcend boundaries—to help broker collaboration and partnerships between schools and families and an array of other institutions that influence the growing child or youth. Arguably, some of the most pressing challenges for educational philanthropy are in the nation's urban areas, where some of the most glaring disparities in resources and opportunities to learn are found and the ideals of democratic schooling are most imperiled. The question is whether current school reforms—choice, single-sex charter schools, market-based efforts, the standards and accountability movement, or the demands of the No Child Left Behind Act, for instance—will marginalize gender issues or see them as an integral part of the process of reconsidering the urban school culture in order to foster excellence and equity.

In sum, one could hardly imagine a more exciting and pivotal time in our ongoing debates about the influences of gender in society and our experiences in educational philanthropy devoted to gender issues. For this reason, the resources that Ginsberg and Gasman bring together here—articles from the 1990s and the diverse perspectives of scholars, grantmakers, and recipients who are currently active in the field—offer a much needed resource for practitioners, scholars, foundations, nonprofit organizations, parents and teachers, and others who give their time, money, and support to projects and initiatives designed to promote gender equity in education and in society at large.

I will conclude as I began while discussing the WCC's ad—namely, by underscoring that questions of gender are if perhaps not always explicit then certainly not far beneath the surface of our discussions about education today; that a consideration of these questions is central rather tangential to our quest to improve education; and that educational philanthropy, today as in the past, has a crucial role to play in promoting access and equity. Rather than an add-on, attention paid to issues of gender in education is consonant with our ongoing efforts to promote diversity and inclusiveness and our expressed desire to provide the best for all children and adults. The types of insights presented in this volume underscore the difficulties of institutional change and systemic reform in education and can help us to improve the efforts and outcomes of educational philanthropy in relation to gender equity, namely by adopting more meaningful approaches to assessment, collaborative and strategic funding, and the task of scaling up of effective projects and reforms. Surprisingly, few educational resources speak directly to the information needs and concerns of project officers and foundation trustees and this volume represents an important effort in this direction. But equally important, by providing an up-to-date "snapshot," if you will, of our gender-related philanthropic efforts and their diverse emphases, the essays that Ginsberg and Gasman have assembled can be used as a vehicle to opening up a more informed and increasingly fruitful conversation among donors, teachers, communities, education reformers, and others about gender in education. Such a dialogue will help advance the work of concerned educators and philanthropists as they strive for social progress and continue to imagine new options and a wider range of possibilities for all individuals, females and males.

Notes

1. A copy of the advertisement is posted on the Women's College Coalition's Web site, www.womenscolleges.org.
2. See, for example, Christine Grumm, "Close Philanthropy Gap by Funding Women," 12 21 03 December 21, 2003, *Women's E News*, and idem, "If Ever There Was a Time, It is Now," Women of Great Kansas City Award Luncheon Keynote Speech, December 2, 2005 available at www.wfgkc.org/htm/Awards_Luncheon05-keynote%20speech.pdf.

References

Havens, J., and Schervish, P. (2003). Why the $41 trillion wealth transfer is still valid: A review of challenges and questions. *The Journal of Gift Planning* 7: 11–15, 47–50.

Rosenberg, R. (1982). *Beyond separate spheres: The intellectual roots of modern feminism*. New Haven: Yale University Press.

Solomon, B. M. (1985). *In the company of educated women*. New Haven: Yale University Press.

Sommer, C. H. (2000). *The war against boys: How misguided feminism is harming our young men*. New York: Simon & Schuster.

Tyack, D. B., and Hansot, E. (1992). *Learning together. A history of coeducation in American public schools*. New York: Russell Sage Foundation.

Acknowledgments

When writing a book, there are always many people to thank. It's a process that cannot be done alone. In our case, we would like to thank our editor at Palgrave, Amanda Johnson, for her support of and belief in this project. She, along with the anonymous reviewers, helped to make the book much better and more tightly focused. We also are grateful to Andrea Walton for writing the foreword to the book. Andrea's work on women and philanthropy, in particular her recent book *Women and Philanthropy in Education* (Indiana University Press, 2005), has inspired both of us. We are appreciative of all the organizations, foundations, and individuals that participated in this project. Their wonderful work was the impetus for our bringing this project together. Lastly, we are thankful for the work that Chris Tudico did on the supplementary bibliography for the book.

—Alice and Marybeth

When I first began thinking of a book on gender and educational philanthropy, it was because, as an independent consultant to foundations, I attended many meetings in which the same questions were raised again and again. These were institutions that in many ways had little in common: some were small in size, some large; some local, some national; some gave large grants, some gave only minor grants; some had been in existence for a long time, while others were just making the way. They all had a commitment to looking at gender in K-12 education, but when the subject came up there was a lot of frustration because they could not decide upon or agree upon how to focus such an initiative, how to convince their board or other stakeholders that this work was important, how to define gender equity, how to target schools and/or individuals to work with, and how to "frame" what they were doing.

I thus want to begin by thanking Shirley Mark at the Schott Foundation, with whom I discussed these issues many times, and who inspired me to pursue them further. On that note, I'd also like to thank Margaret Hempel at the Ms. Foundation for Women, with whom I also spoke quite a bit about gender and educational philanthropy, and pointed me to Ms's publications, which were compelling and groundbreaking.

Of course I want to thank my coauthor Marybeth Gasman with whom it has been my greatest pleasure to work. To this day it astounds me how much we think alike despite the fact that we come from two very different areas of education learning. Marybeth has taught me a great deal about both philanthropy and the world of publishing. She is both dynamic and reliable, a leading intellectual and a compassionate person. I thank her greatly.

I would like to thank friends and colleagues—Rachel Allender, Shirley Brown, and Joan Shapiro. Finally, I would like to thank my family. My mother Lois Ginsberg has cornered the market on generosity. My two sons Nicholas (7) and Andrew (11) remind me daily why it is important to genuinely include boys in our investigations of gender. Finally, to my husband Sam, I can only say that, as with my previous book, this book could never have been written without him. As we approach our twentieth anniversary, I find my love for him growing stronger each day and only hope that I can support him as much as he has supported me. Indeed, I dedicate this book to my husband with everlasting gratitude.

—Alice Ginsberg

When Alice approached me in 2004 with the idea of coediting a book on gender and philanthropy, I was intrigued. Although most of my research is related to philanthropy, I had done little related to gender. My work primarily focuses on African American education. Editing this book with Alice has proved to be a wonderful learning experience, one in which I have greatly expanded my knowledge of the larger field of philanthropy. I am grateful to Alice for bringing me into this project. She is a terrific scholar who is committed to making positive change in our nation's schools and in the lives of boys and girls.

I would like to personally thank Andrea Walton for her support of my career and for her contributions to the field of philanthropy. I am also indebted to my research assistants at the University of Pennsylvania who support my ideas on a regular basis. Thank you to Chris Tudico, Noah Drezner, Katherine Sedgwick, and Shannon Gary. I am blessed with wonderful colleagues at Penn and I thank them for their support of me and my research: Matthew Hartley, Laura Perna, and Susan Yoon. Lastly, I must thank my family. My husband, Edward Epstein, is a constant source of support and inspiration to me personally and professionally. Likewise, my sweet little girl, Chloe, is the best daughter a mother/professor could have. I thank her for always making me laugh hysterically!

—Marybeth Gasman

Introduction

Alice Ginsberg and Marybeth Gasman

From Women to Girls

That philanthropic organizations recognize women's issues as a specific "category" of funding is not new. As early as 1972, the Ms. Foundation for Women, which presented itself as being out of the mainstream, became one of the first funders to support domestic violence issues. Subsequently, this organization supported initiatives for women's economic empowerment, reproductive rights and the eradication of workplace discrimination. Over the years, smaller women's foundations grew across the country and many of the programs that Ms. pioneered are now, in the foundation's own words, "replicated by mainstream funders" (www.ms.foundation.org). In addition, a number of umbrella organizations such as *Women and Philanthropy* have been founded and developed to publicize and coordinate these activities.

Gradually, questions about *girls* were raised in conjunction with those about women. For example, in *A Conversation About Girls*, a series of conference papers conducted throughout the 1990s (and included herein), the Valentine Foundation and the nonprofit Philadelphia-based organization Women's Way raised a number of critical questions, including the following:

- What knowledge must funders acquire to support girls in an intelligent way?
- How does positive self-esteem help girls to develop their "own voice"?
- How do racial, economic, and ethnic factors affect the process of finding one's voice?
- What kinds of programs can promote finding one's voice?

Such questions led to significant research, an interest in addressing bias against girls in school and community settings, and in encouraging funding agencies and policymakers to give more attention to girl-specific programs.

On the other hand, the idea that gender might be a separate funding category is much more recent, and has been met with considerably more controversy, confusion and, in some cases, even curiosity. Whereas funding for women and girls grew directly out of the field of women's studies and feminist discourse, funding for gender is significantly more complex. Foundations must ask questions such as: Is gender used primarily as a synonym for girls? How do girls and girls' issues differ from boys and boys' issues? Can we make such generalizations, given the many other ways in which both boys and girls identify themselves? Perhaps, most importantly, in what *context(s)* do gender variations make the most difference? One such context is clearly education. In this book we will examine the relationship between philanthropy and gender in the arena of K-12 education, particularly, though not exclusively, around urban school reform.

In the 1990s, the publication and dissemination of reports and books such as the American Association of University Women's (1992) *How Schools Shortchange Girls: A Study of Major Findings on Girls and Education* and Sadker and Sadker's (1994) *Failing at Fairness: How America's School Cheat Girls* brought widespread attention to issues of gender bias (albeit primarily against white middle-class girls) in American schools. These authors named a variety of ways in which girls were not treated fairly in schools, ranging from: being subject to sexual harassment; not being encouraged to pursue higher-level courses; not being directed to careers in math and science; not seeing themselves represented in the curriculum; and having teachers pay more attention to boys (it was found, for example, that teachers called on boys more often even when they did not raise their hands).

These publications, along with others that quickly followed, were based on new and significant research conducted in schools, many with the full support of teachers and administrators. The studies underscored the ways in which girls and boys can be both differentially and disproportionately affected by educational and social norms. They also underscored the need to look at gender across the curriculum, not just in language arts and social studies, but in subjects such as math, science, and art. They called into question not only whether women and girls were adequately represented in the curriculum, but also where, why, and most importantly how they were represented, or diminished. In this sense, the studies advocated a curriculum transformation that went beyond "add women and stir." The curriculum became a vehicle for asking questions about the definition of American culture and identity—questions that we will soon argue are highly relevant for boys and girls alike.

Moreover, this research corresponded with a new era of *professional development*, one in which teacher inquiry communities, action research, and ongoing collaboration was promoted over teacher "training." In this environment many teachers were given more authority to create and build upon curriculum themselves. Of particular importance, teachers attempted to develop a curriculum that more fully engaged students by drawing upon their personal experiences, cultural identities, and other issues dominant in the communities in which students lived.

In the early 1990s many public school teachers were, for a short time, less pressured to teach packaged or scripted texts—although it must be acknowledged that current educational policy/reforms, such as the No Child Left Behind Act (NCLB) have certainly challenged this ideology. NCLB, though well intentioned, is particularly troubling because despite its democratic language it does not address real issues of discrimination or bias, or equip students and educators with the tools for fighting it. Its primary focus is on standardized tests, a great many of which do not reflect students' true abilities to contextualize and use knowledge, and certainly do not provide extra resources and suggest explicit ways that will ultimately make schools more equitable.

Nonetheless, we see many examples of teachers and administrators who are aware of gender issues in their classrooms and who find ways to raise the kind of compelling questions—even if these questions are not "on the test." Many programs, such as the long-running and nationally recognized SEED (Seeking Educational Equity and Diversity) Program, and the National Writing Project are devoted to promoting these networks. Many of these educators, working side-by-side with community education institutions (such as libraries, museums, after-school and community centers) have developed compelling programs to address *what is not happening* in the classroom.

For example, in a project called GATE (Gender Awareness through Education) middle and high school teachers across the curriculum posed the same question to their students: "What is women's work?" Students then explored the issue from a multiplicity of perspectives by reading a range of literature on women's work in different cultures and historical eras—then actually going out into their communities and interviewing family members and others about women's work lives.

As a result of this project, many students commented that they never realized how hard their mothers worked, while others questioned whether staying home and taking care of the house and kids was to be considered as work at all. The teacher, in this case, followed up with an exercise asking students to add up the financial value of what their mothers' did if they had to pay someone to do the various tasks. This certainly shed a new light on women's work, and at the same time crossed disciplines from history to math. It is important to point out that this exercise was as compelling for boys as it was for girls—like girls, many boys did not appreciate or value their mother's work until they were actively exploring and questioning it. This is, no doubt, part of the philosophy of including boys in Ms.'s "bring your daughter (and now sons too) to work" day.

So What about the Boys?

David Sadker has written that if you bring up the issue of gender equity at a social gathering, you are likely to hear one of the following responses: "Gender equity—we dealt with that years ago [or] Isn't it really boys that have problems today?" These assumptions are deep-rooted, and usually not based on research, but rather poorly researched media messages, "campaigns" for boys (such as that of First Lady Laura Bush), and the faulty assumption that if we have been helping girls all these years we must therefore have been "hurting" (for example, depriving) boys.

We are greatly pleased to report that there is an increasing body of literature focusing on issues of gender socialization and masculinity, and helping us to better understand boys' problems in schools and their surrounding communities; moreover, much of this literature focuses on boys of color, from diverse ethnic communities, and from working-class communities. Key findings of this research were that boys are just as constricted by narrow definitions of masculinity as girls were of femininity (for example, that boys could not show any signs of doubt or weakness); that boys are not encouraged in disciplines such as language arts, and when they are, teachers may overlook the areas where a different teaching approach is necessary; that it was "not cool" for boys to like school, and that much of what was taught in school did not address the very serious issues of violence these boys faced on a daily basis in their schools, and most especially in their communities.

We also want to say from the beginning of this venture that we do not believe in setting up boys and girls in a dichotomous relationship, where helping one means shortchanging the other. Quite the contrary: in order to achieve equity, respect, and healthy relationships between boys and girls (and later men and women) programs designed to help one, should, ideally, also be constructive for the other.

What is Gender Equity in Education?: Different Approaches

While the idea that what helps girls should also be helpful for boys and vice versa seems relatively straightforward, it is far from simple for foundations and philanthropists looking to fund, prioritize, and support gender equity programs in education. Education funders have to make difficult decisions about how to use limited resources. As previously noted, they must grapple with questions such as: what exactly is meant by the term equity? Several different and widely divergent approaches have been identified here.

One approach is something akin to being "gender-blind." This means, ideally, treating boys and girls alike. Teachers and other educators strive to "not see gender" (and in most cases, other areas of difference among students). In this approach teachers would simplistically aim to call on students of both genders exactly the same number of times, and they would provide consistently similar feedback and promote activities that did not emphasize one gender over the other. Assignments and textbooks would include the experiences and accomplishments of women and men in equal numbers.

There are many problems with this approach, including the fact that teachers often have unrealized biases of their own, and thus continue to treat boys and girls differently even when they are trying not to. Another potential drawback is that this approach is often reduced to—"add women and stir," where women are simply inserted into the curriculum for the sake of equity, without ever asking students to think critically about why and how they were excluded in the first place. Thus instead of initiating open conversations about identity and equity within the classroom, they are, essentially, being shut down.

Another approach is teaching to students' (real or perceived) differences. In this case, teachers and other educators assume that boys and girls are inherently different—whether through socialization, through biology, or some combination of the two—and thus need to be taught in different ways. An example might be that girls learn better cooperatively and should be given opportunities to work in small groups, while boys work better competitively and should work more individually.

One of the potential problems of this approach, however, is that it assumes that boys and girls are dichotomous categories, when in fact, not only do differences exist among girls and boys, but, taken a step further, each student is unique, and it is critical to look at issues of race and class, sexuality, religion, and other differences besides gender. Foundations, in fact, constantly grapple with the question of whether all boys and girls are equally disadvantaged and should be targeted equally, or whether they should be paying special attention to certain groups of children within gendered categories—such as African American boys who are often regarded to be the most "at-risk" gender group (Dance, 2002; Finn, 1999; Noguera, 2003, 2005; Sewell 1998).

Yet another approach is likened to "affirmative action" where many teachers feel that resources (used here in the broadest sense) have been unfairly distributed to primarily white, middle-class boys and therefore must be redistributed. In its most basic interpretation, this is the intent of Title IX (currently under attack by policymakers), which mandates that girls' sports get equal funding and attention as boys' sports. Likewise, teachers may set up special single-sex support groups for girls who do not speak much in class, to give them an opportunity to voice their opinions in a "safer" atmosphere.

Again, this approach has its pros and cons, the most obvious problem being that it pits girls' needs against boys' needs rather than recognizing that what is good for one is likely good for the other, as the ultimate goal is to build healthy relationships and promote academic achievement for both groups. This attitude is difficult to change, because our school system is set up as a meritocracy, and, because realistically girls and boys must compete for the same share of limited resources—especially in poor urban schools.

This is not to say that there is not great value in these three approaches. Rather, we are making the case that they must be considered together, and, moreover, in the context of ongoing professional training and development for educators who are raising questions about gender overtly and internally all the time.

Youth as Grantmakers and Evaluators

Another issue that philanthropists interested in gender and youth programs face is how to get youth involved and invested in specific programs. This is important because many gender-equity education programs—whether focused on curriculum, school climate, or after-school programs—are developed by adults. This would not be a problem per se, but many teachers never addressed gender directly in their pre-service work, and are not aware of the ways in which youth think about gender and the kinds of critical questions that need to be asked continually.

A number of foundations are beginning to address this by creating philanthropic initiatives that are organized and run by youth (see part two of this book). In such programs, participating girls (and ideally boys as well) may be given the opportunity to develop a request for proposals (RFP), thus deciding categorically how money will be spent and resources will be distributed, and subsequently deciding specifically which groups will receive funding and/or resources. Likewise, an increasing number of young people are becoming involved in the evaluation of programs, usually in conjunction with adults, which gives them an opportunity to ask critical questions as to whether the program achieved its goals and how important these goals turned out to be. The bottom line, however, is this: youth need to have a voice in the funding process.

Such evaluations should be quantitative and qualitative, participatory (that is, include a range of stakeholders), internally and externally driven, and, perhaps most importantly, value ongoing inquiry as a way of improving the services these programs deliver. It should be noted that many of these new kinds of assessment—such as an entire tool kit developed by the Ms. Foundation for Women—are written with specific gender-focused questions in mind (see Ginsberg's chapter [chapter twelve] of this volume).

As schools are mired in policy mandates such as No Child Left Behind that rely almost entirely on standardized testing, it is difficult for funders to support and/or accept evaluations of programs that are not primarily quantitative in nature. More qualitative and anecdotal evaluations are looked on with some skepticism because they don't seem to "prove" that students are increasing their academic achievement. This is a serious and ongoing problem for funders. The issue of accountability is critical, but the questions remain: accountable to whom, by whom, and for what purpose? This means that funders need to understand not only how many people attended a program, or even how much they liked it, but also whether or not the experience provided a new understanding of a problem or subject, and whether the program influenced their future behavior and attitudes.

Such evaluations also point to the need for grantmakers to fund programs over a period of time (for example, to give programs time to evolve and produce meaningful results), as well as to have additional funds specifically earmarked for evaluation of programs. Some foundations have even adopted a "cluster review" approach, wherein they examine entire grantmaking areas rather than focusing on one or two "failed grants." This approach fosters a learning environment where the emphasis is not so much on "mistakes" but on a better understanding of, and increasing creditability for the philanthropic field as a whole.

Collaborative Funding

One other area that this book addresses directly is the idea that funding individual initiatives to promote gender equity, though very important, is not always the best strategy. This is true for a number of reasons: (1) when the funding runs out, many of these programs simply fade away; (2) grantees do not have the opportunity to learn from each other and thus, often, reinvent the wheel; and (3) when philanthropists work collaboratively, it provides an opportunity to promote systemic change in a school, school system, academic discipline, or area of concern, rather than a smattering of short-lived individual programs.

Such collaborations also change the dynamics of the relationship between "benefactor" and "client," as it is recognized that each can learn from the other, and that funds spent on one project may provide valuable lessons or serve as a model for other projects in similar contexts or demographic communities. Moreover, assessment of programs becomes less of a show-and-tell as there is greater opportunity for *honest* feedback. When parts of programs do not go as planned they are no longer labeled "failures," but, as noted, are viewed as important learning experiences for all stakeholders.

The Kellogg Foundation and the Annie E. Casey Foundation, for example, have regular meetings between evaluators and program staff five or six times a year. The James Irvine Foundation has created role of "learning coaches"—external agents supported by the foundation—to convene program staff, evaluators, and multiple grantees in order to discuss specific questions or issues they share. In part two we include an essay by the former Gender Equity Program Manager at the Schott Foundation, Shirley Mark, which, through its Gender Healthy/Respectful Schools, brought funders and grantees together on a regular basis to hear about and learn from each other's programs, and to brainstorm ways in which they could achieve strategic change (for example, policy changes).

Having youth participate in the funding process, and bringing funders and grantees together are both critically important parts of developing new models of fundings as they each underscore *the value of looking at gender equity from the bottom-up rather than from the top-down*. Both sections of this book strive to illustrate new kinds of philanthropic processes that explore and challenge definitions of gender equity in education and that encourage grantors and grantees to work together in new and innovative ways.

Organization of the Book

As suggested above, funders interested in supporting gender equity work in schools face many challenges. This book addresses these challenges in the following two sections:

Part One: Definitions of Gender Equity

Defining gender bias and gender equity is not at all straightforward; we can no longer define boys and girls as dichotomous categories. In other words, how do you do grantmaking with a "gender lens"? How do foundations use funding initiatives to explore relationships among the categories of gender, race, ethnicity, class, and sexuality? How do we come to see and understand that gender is not a singular identity, but part of a multifaceted identity or identities?

The first seven chapters of this book (part one) address these issues head-on. For example, in "Talking about Gender Equity and Education: A FrameWorks Message Memo"—a study, commissioned by the Caroline and Sigmund Schott Foundation (Cambridge, MA)—the authors surveyed a cross-section of people about how they "frame" gender issues in schools. This was an important undertaking for the foundation, and for other foundations interested in the same topic, because unless we understand how people think about gender equity it is difficult to craft a funding initiative that will be taken seriously and will be long-lasting.

The authors found several pervasive attitudes about gender in education, including the following: that people have a strong desire to see the classroom as a "controlled environment where children are protected from social problems, not exposed to them." They refer to this as the "consumer model" of education, where parents are particularly interested in their own child's achievement more than in broader issues of social justice.

This "frame" for thinking about gender equity programs in school and in after-school programs is pervasive among funders themselves, and is strongly echoed in Molly Mead's work, "Gender Matters" (National Council of Research on Women), which suggests that "[t]he bottom line is that funders have a strong preference for funding so-called universal (or coeducational) programs, with little awareness of the need to consider gender when setting grantmaking priorities or allocating funds to grantees" (see chapter two of this volume).

She explores funders' assumptions about efficiency (for example, "we have limited resources"); democracy (for example, "targeted programs promote exclusivity"); and relevance (for example, "gender is an irrelevant category for targeting").

Mead suggests that much grantmaking is political in nature, and that when funders consideration gender in their grantmaking, they often do so because it is relevant to their funding priorities as opposed to a concern for equity or fairness. The authors conclude, however, that the biggest obstacle to gender-focused funding is that "foundation executives, trustees, and program officers lack knowledge or understanding of gender differences and do not acknowledge that there can be disparate needs based on gender."

In chapter three, "Grantmaking with a Gender Lens" (Grant Craft, a project of the Ford Foundation) the authors try to sort out "what gender analysis is and isn't." In particular, they note that "gender analysis doesn't compromise neutrality," nor does it apply *only* to girls and women. The chapter directly addresses the need to include men and boys in gender programming.

Chapter four, "Bringing Funders Together to Talk about Girls: A Roundtable Discussion," documents a conversation about gender and funding that took place over fifteen years ago. This chapter is the result of a collaboration between the Valentine Foundation and Women's Way (both in Philadelphia) in which funders, academics, policymakers, and girls' advocates were brought together to address their belief that "funders lack opportunities to hear researchers and advocates talk about the very real unmet needs of girls (p. 90)." One can see that many of the issues raised in this chapter continue to be relevant, even if some of the approaches to further defining and addressing them have changed.

The issues raised were as follows: Issues of intersections among race, class and gender, for example, were raised as it was noted that middle-class white girls' have different experiences than girls of color and working-class girls. The second issue raised herein is the issue of "exploring the possibilities of comprehensive, sequential, sustained collaborative funding among foundations (p. 92)." The importance of listening to girls' voices is also raised as they noted, "We must not let our own ageism interfere with the design of programs that may be more effective than those we ourselves design." Also of note in this chapter is the idea that "[m]entoring programs should provide training for the mentors and ongoing evaluation of the program." As noted herein, although this piece is dated, it is included both for historical reasons and because many of the issues and questions it raises are still on the table today.

Rosa Smith does exactly this in chapter five, "Saving Black Boys," which specifically focuses on black boys. Smith underscores the risk black boys face and suggests the importance of raising "awareness and support strategies" to redirect the K-12 public school achievement trajectory of black boys. This includes "building partnerships and convening quality public meetings to listen to concerns and recommendations of students." According to Smith, twice a year, this process should link funders specifically with black male students.

In chapter six, Michael Reichart similarly notes, "There has been a strong reaction to the 'me too' nature of the men's rights theorists and their unwillingness to address dynamics of power and male privilege relative to revision (p. 119)." Instead of pitting boys and girls needs against each other, however, Reichart argues that boys are stuck in rigid, hegemonic notions of masculinity and much of this is both tolerated and even encouraged in the "boys will be boys" attitude but that "much is changing on the gender landscape" and that mentoring and other programs designed to raise boys self-esteem can be critical. According to Reichart, saying it is great to be a boy does not have to be the equivalent of saying that "being male is better than being a girl."

Similarly, in the Ms. Foundation's chapter (chapter seven) on supporting boys resilience, the foundation brings together academics, the media, community planners, and service providers to look at the ways in which masculinity and femininity have been "culturally constructed" and the ways in which this constricts boys' lives—albeit in different ways than girls.

The report looks at some perceived obstacles to boys' achievement in schools and compares them to real obstacles. It specifically addresses issues of race and class, noting that all boys are the same, and foundation should not treat them as such, and therefore raises the question: "What does it mean to be asked to give up power when you feel powerless?" It concludes, "Men's relationship to gender-based domination must be located within the context of the social justices that structure society for women and men alike (p. 152)."

Part Two: Collaborations and Program Assessment

As noted above, an exploration of gender and educational philanthropy invites many different kinds of collaborations, including having youth themselves participate in the funding and evaluation process, and arranging, on a regular basis, meetings of donors, donors and grantees, and groups of grantees. The chapters in this part explore these issues in more depth.

In chapter eight, "Power and Possibilities: Collaborative Fund for Youth-Led Social Change," the Ms. Foundation explores its innovative program to involve youth (in this case girls-only) in the process of philanthropy by giving them the power to make decisions about where money is targeted. It is part of a larger collaboration among funders, youth organization staff, and youth leaders, in which "the partners work together to shape initiative activities and information-sharing opportunities." This includes new models of mentoring and a "genuine commitment to two-way learning where youth and adults work with each other, learn from each other, and shape leadership roles."

In chapter nine, "Young Women for Change," the Michigan Women's Foundation leads a similar program in which it has a "girls-as-grantmakers" program in which high school girls of diverse backgrounds assess the needs in the community and then make grants that they believe will fulfill those needs. Among the program's overarching goals is to "prepare a new generation of young women for the reins of leadership in philanthropic endeavors." As one alumni of the program remarked, "I learned how to give wisely. . . . I learned how to be a team player. . . . We were opinionated. We worked well together in spite of our differences (p. 177)." Another remarked, "I became keenly aware not only of the problems facing young women, but the organizations that are making it their mission to find solutions, inspire, and empower those who need assistance."

Chapter ten is an interview between the book's editors and a former program manager at the Schott Foundation, Shirley Mark. Mark was instrumental in developing the foundations' program Gender Healthy/Respectful Schools, which was remarkable in that it was not only one of the first grantmaking programs to focus specifically on gender in K-12 education, but brought the directors of program grantees together on a regular basis to share ideas, questions, experiences both with each other and with the foundation. In this way, the grantees were able to learn from each other, not to reinvent the wheel, and together work toward larger policy-scale changes. It also gave the funder a much closer look at what they were funding and whether or not the programs were meeting the funders' goals.

Chapter eleven, "Sisters Empowering Sisters and a Case Study: Girl World" is authored by The Girl's Best Friend Foundation (GBF). As with some of the other grantmaking programs that put youth in charge, Sisters Empowering Sisters, a program of the Girl's Best Friend Foundation, recruits up to 15 young women each year to participate. The program which includes (1) activities and discussions centered on grantmaking, social justice, and sisterhood; (2) evaluation of grant applications and decision making about granting money to projects developed by girls for girls (ages 8–18); and (3) meeting and sharing information with GBF board, staff, and other groups of young women in the Chicago area and across the country.

Important questions are raised in this process: What do youth have to say about gender issues and how do they prioritize the need for a more gender-equitable education? Are there differences among the ways in which different groups of boys and girls view their experiences in school? What factors contribute to these differences? According to the authors, "We are going past paper and print to see the passion for and dedication to their project in person (p. 191)." Girls are also extensively involved in the evaluation of these programs, and the group has developed some very impressive guidelines for doing so. This chapter also includes a case study of one program that focuses on girls investigating and evaluating the results of their own work through interviewing, photo journaling, observation, focus groups, and the like. Such case studies bring out the vigor of this kind of work. They also point to the obvious fact that if programs are going to change, then the mechanisms by which they are evaluated and assessed must also change.

There is a need for new kinds of (process-based) evaluations and assessments that are specifically designed to improve gender-focused programs. Yet questions arise: Who should be involved in the assessment process, and to whom are we most accountable? What is the role of youth in the actual evaluation process? What kinds of qualitative data can be used and what does it tell us? How can evaluations be done in such a way as to foster a learning environment, where changes in programs, and "mistakes" can be seen as opportunities rather than causes for blame.

In chapter twelve, "Gender Equity in Urban Education: New Relationships between Funding and Evaluation," Ginsberg considers some of the common obstacles involved in evaluating gender equity programs in schools and community organizations, especially as most of them cannot be evaluated through standardized testing. It is argued that even those funding agencies that are already committed to supporting gender equity programs have a difficult time targeting coveted resources for what are perceived to be "specialty" programs, benefiting only a small number of students, and disconnected from larger school reform goals surrounding student achievement.

The ideas presented in this book, though in some instances written many years earlier, are largely echoed in a recent report by the Girls

Coalition of Greater Boston (written 2005, released 2006) in which they "sent out a call" to researchers, funders, leaders of youth programs, and policymakers to take a more active role in girls' after-school programs. In regards to funders, they conclude with the following recommendations, *among others*:

- Give gender-sensitive programs time to develop and succeed (multiyear funding);
- Provide funds for evaluation, especially participatory evaluation involving input from girls (and boys);
- Provide funds for professional development (for example, youth-worker certification, ongoing gender-sensitivity training); and
- Facilitate collaboration and resource sharing among gender-sensitive programs.

As we have noted several times herein, this report also underscores that one of the key problems facing gender equity in education is the "perception that the problem has been solved, and the mentality that if girls win boys lose." Yet as they so rightly note: "Neither boys nor girls benefit when their needs are pitted against each other."

We conclude by noting that although this book has gender as its primary focus, many of the questions raised about philanthropy and education are enduring, and can be useful to think about in other contexts as well. We want to underscore, once again, our belief that girls and boys are not dichotomous categories, and that within each gender there is great variability—much of it based on race, ethnicity, and class. This is not easy work—either for funders or grantees. Yet we hope the book will highlight, in the long run, the kinds of questions and strategies raised that need to be pursued if our ultimate goal is, in fact, gender equity in education—regardless of which definition(s) we use to define it.

Part I
Definitions of Gender Equity

Chapter One

Talking about Gender Equity and Education: A FrameWorks Message Memo

Joseph Grady and Axel Aubrun

This work has been undertaken as part of the Caroline and Sigmund Schott Foundation's ongoing enterprise to promote gender equity, and more specifically, as part of the current effort to promote gender equity in the context of education. In particular, we have set out here to determine how American adults concerned with education understand the nature of the relationship between gender and the school experience (as well as related identity issues such as race and class). We focus not on opinions about particular issues relating to gender and schools (some of which have been documented in previous polls and surveys), but on the more fundamental cultural models that define the key concepts associated with School and Gender, and on which the various opinions are based. A clear understanding of the public's deeply held and widely shared assumptions will be instrumental in helping foundations shape effective communications campaigns.

Method

This chapter is based on a series of twenty in-depth, one-on-one interviews—or "elicitations"—conducted by Cultural Logic with parents, educators, administrators, and other adults in Massachusetts (primarily Boston and Cambridge, but also Western Massachusetts) concerned with the issue of education. Subjects came from a range of ethnic, cultural, and economic backgrounds, and included a number of self-described conservatives.

"Cultural Models Research" (CMR)—based on cultural anthropology and cognitive linguistics—provides an unusually detailed picture of people's deeply held and usually hidden assumptions about a given topic. And unlike traditional qualitative research techniques that are typically aimed at discovering stand-alone opinions, CMR focuses on the *connections* between basic assumptions.

Understanding these connections is the key to understanding how people reason about a given topic, for predicting how the public will respond to new ideas, and for suggesting new ways to frame issues. In summary, CMR is designed to provide a clear picture of strong, recurrent patterns in thinking—in this case a set of values and beliefs that underlie a distinctive understanding of the relationship between school and gender.

Summary of Findings

Many people feel there is little or no gender inequity in the classroom—the classroom is seen as set apart from society. Simply put, gender equity in schools is not seen as an important problem by most people, even when they agree that there are gender inequities in society more generally. This may reflect the fact that an important part of the American cultural model of schools holds that they are "set apart" from the larger society, safe and controlled environments in which children are protected from the evils of the world.

If gender equity in schools is a meaningful issue, it isn't reducible to "helping girls succeed." Considerations of gender equity are informed by a more basic model of "fairness," which in this case encourages Americans to consider the points of view of boys as well as girls.

One important model of gender inequity holds (in its strong form) that boys are "defective" while girls are "disadvantaged." When gender inequity is accepted by Americans, it is sometimes interpreted as the idea that boys have internal problems (intellectual deficits, attention problems, propensity toward aggression, inability to express feelings) and external advantages (prejudices in their favor, et cetera) while girls are internally gifted (morally, intellectually, socially, and emotionally superior) but held back by external circumstances. An implication of this model is that boys need to be constrained while girls need to be liberated.

An alternative to the "Defective Boy" view is the "Bad Fit" view. Many people believe that the way schools are traditionally structured seems designed to have boys fail. Boys are seen as more active and less able to adapt to the constrained setting of traditional schools, and for that reason more likely to be treated as troublemakers.

People are sometimes conflicted on the issue of gender equity in the classroom because of a clash between "public discourse" and "lived experience" on this issue. Many people seem to experience something like "double-think," simultaneously believing that girls are discriminated against (a model that is prominent in public discourse), and that teachers and the school setting favor girls in a number of ways (a view that is often more consistent with their personal experience).

There are two conflicting ideas of the ideal classroom: one emphasizing individual students and teachers and one emphasizing the "culture" of the classroom. The first view leads to a foregrounding of public policy issues such as the importance of "individualized instruction," with an emphasis on taking account of differences between students and between teachers and students. The second places the emphasis on a shared classroom culture rather than on the individual qualities of teachers and students.

Key elements of social/emotional experience are seen as playing a critical role in academic success. While lay people sometimes fall back on a simplistic separation between academic performance and other aspects of a child's personal experience in school, they more often demonstrate an awareness of the strong causal connection between these two domains. The issue of gender equity plays out differently at different age/grade levels. When people think of "school," they tend to think first of elementary school—a space that is largely protected from the dangers of the larger world, including its gender and racial inequities. According to this model, high school, by contrast, is much less separate from the influences and problems of the larger society.

Most people believe that there are important differences in how teachers should communicate with different individual children, but not different groups (for example, ethnicities). An important part of the American model of education is the idea that teachers must tailor communication to individual children. There is little emphasis on the importance of tailoring the communication process to different groups—perhaps due to the contradiction with the idea that one purpose of school is to forge a common identity.

People have a consumer model of education: it's mostly about getting the best quality and best advantage for their own kids. People feel that ideally their kids will be engaged and excited by school. But the bottom line is that the school system should ensure that their children will have all the advantages that other kids have.

Findings Related to Specific Language

In addition to findings about relevant American models of education, the interviews revealed consistent attitudes toward two phrases that are prominent in the public discourse on education:

Gender equity. This phrase is not easily understood by most people, and when discussed in terms of the specific disadvantages faced by girls, it sometimes leads to resistance.

Individualized education. While this term conjures an ideal which most people support, it often has slightly utopian connotations, amounting to an idea that is wonderful but impractical.

Discussion of Findings

Many people feel there is little or no gender inequity in the classroom—the classroom is seen as set apart from society. While it is widely acknowledged that women do not enjoy the same degree of opportunity for success in society as men, subjects do not necessarily agree that this disparity extends to the classroom. The majority of subjects feel that gender inequity in the classroom is probably inconsequential, and that where it does exist it is the product of a particular teacher or school rather than a systemic problem.

In effect, the classroom is often seen as a sort of (permeable) bubble, set apart from the outside world. Social and political issues that affect society as a whole are largely absent from most people's view of the prototypical classroom. This is not to say that the classroom is thought of as entirely free of society's inequalities and prejudices; some negative aspects of society do "leak" into the classroom—for example, some teachers are certainly racists (usually in subtle ways); class, family situations, and economic status do affect students' experience and success in school, and so forth.

While teachers are the most obvious agents who bring aspects of society into the classroom, students themselves also carry in some baggage from the outside world. In particular, the child's family experience—whether the child comes from an encouraging family or not, the degree to which the family is engaged in the child's education, et cetera—is an important aspect of the outside world that has an impact in the classroom. In addition, negative messages that children have internalized from the media—that girls are not as capable as boys in certain areas—are also perceived as finding their way into the school experience.

Another sense in which schools represent a "bubble" is that they are often seen as having resisted historical change. That is, many lay people do not feel that schools or the school experience have changed much over the decades. (Not surprisingly, experts tend to see the situation differently from most nonspecialists. They tend to be aware of gender-based differences in boys' and girls' experiences in school, though this awareness can take several different forms, as we discuss in sections below.)

The strategic implication of this finding is that the Schott Foundation certainly cannot assume that the public will be moved by a campaign built on redressing inequities. For most people, the inequities simply are not there.

If gender equity in schools is a meaningful issue, it isn't reducible to "helping girls succeed." Considerations of gender equity are informed by a more basic model of "fairness," which in this case encourages Americans to consider the points of both girls and boys.

Almost no one felt girls were disadvantaged in the classroom (but see attention issue, below). In fact, from early on, school is felt to be a better fit for most girls than most boys (see below). It is boys who have troubles with school: trouble sitting still, reading, expressing themselves, following rules, quelling aggressiveness and competitiveness. Even through their teen years, when children of both genders are concerned with image, girls fit schools more naturally. The dominant, shared cultural model of girls predisposes them for compliance and success in the classroom, whereas the comparable model of boys does not.

Attention from teachers is an important issue, however. Many people feel that boys get more attention than girls (often for inappropriate reasons), with the result that girls are relatively neglected. On the other hand, others believe girls get a proportionately large amount of teacher attention, based on the fact that girls are better students and thus that teachers prefer them.

> [T]he fact that boys acting up get more of your attention . . . much more than their share of teacher's time. I do think that boys, because of their greater aggressiveness for whatever reason, tend to take and get more of the attention in schools.

The most easily accepted conception of gender equity is closer to "gender-blindness," than to "affirmative action for girls." A gender-blind education would be one in which teachers (and school policies) would work toward distributing attention and support fairly; toward implementing an unbiased curriculum; toward teaching mutual respect and tolerance; toward up-ending expectations of success or failure in certain areas according to gender (for example, girls in math and science; boys in expressing themselves or "behaving"). If schools were gender-blind, everyone would be able to achieve their full potential, without biases and preconceptions getting in the way.

Though the model of racial and class equity is less salient in people's minds, the evidence suggests that a similar kind of reasoning would apply to these kinds of equity as well—which indicates a specific challenge for efforts to communicate the importance of cultural diversity, for example.

> Boys couldn't play a violin, and girls couldn't hit a home run. I think kids do that to each other as soon as they get exposed to TV and social stuff.... [G]irls have an image, but it's an image where you're not supposed to be acting out in school, and you know, you're there to learn ... boys ... feel that they have this image that they have to keep up, and like, well, I don't want to be like oh sits there and he's all into his books, and he's nerdy, and you know and all this and you got this other one that wants to be the class clown, because he like I said he doesn't know how to reach out and ask for help.

One important model of gender inequity holds (in its strong form) that boys are defective while girls are disadvantaged. While most people are hardly aware of the idea that there could be significant gender inequity in schools, or else feel that there are unnecessary pressures on both boys and girls to conform to stereotyped expectations, there is also a view, held by a smaller but vocal group, that gender inequity is important and asymmetrical. In effect, this group believes that boys suffer from internal inadequacies (and are advantaged by external circumstances), while girls are largely the direct or indirect victims of external circumstances (including boys' behavior).

In its extreme form, this "Disadvantaged Girls/Defective Boys" model casts boys as defective and even immoral. According to this model, boys are inferior to girls in various ways, and a noxious influence, and it is important to gain greater control over them in the classroom.

> I think girls are more intelligent. You know, ... it's just my bias. But that's what I think.... [G]irls have more ability to speak, ... express themselves, to think, to organize information, to keep everything on track, keep the group together.... If I have a group of all boys, it will be difficult for that group. They will have chaos and mayhem.

According to this view, the greatest problems facing girls are largely imposed by their surroundings (unlike boys' problems, which belong to their nature as boys). They suffer particularly from not receiving enough attention, and from not being held to the same high standards—for instance, from being expected to perform more poorly.

The appropriate remedies implied by this view would seem to involve greater control of boys, by various means. "Strict Fathering," for instance, in the sense discussed by George Lakoff in his book *Moral Politics*, involves behavior management through forceful discipline. And an option often brought up in the context of boys' "problematic" behavior, though not by subjects in this study, is the hyperactivity drug Ritalin.

> [Boys] always want to show which one is the toughest, or if there is a new student the other students like to show which one controls the classroom. One of them must start something just to let the new student know that they are in control.

There are, of course, much milder or partial forms of this view—for instance, many people express the idea that girls are more focused and capable in school, or that boys get into trouble more, without going so far as to imply that boys are "bad."

Interestingly, this view is encountered more often among education professionals than among other adults. While this may be because their greater experience tells them that there are real differences between boys and girls in school—differences which the public at large is much less aware of—a caveat is still warranted: Framing boys' difficulties in school (which are very well documented and quantified) as consequences of innate shortcomings is not necessarily justified (see below).

While the view that girls are shortchanged in school has currency among some populations, there are strong and potentially counterproductive versions of this view which foundations should be careful to avoid in their efforts. An alternative model, potentially the basis for more constructive reframing, is discussed in the next section. An alternative to the "Defective Boy" view is the "Bad Fit" view. Many people believe on some level that school is not very natural for kids in general. The setting works against many of children's natural tendencies rather than playing to their strengths, and it takes a great deal of effort and skill on the part of all concerned to make the system work well. (Very good schools, of course, can actually feel like natural places for kids, but this is more the exception than the rule.)

One way in which this question of naturalness comes up is that avoiding boredom and "keeping kids interested" play very salient roles in people's understanding of the classroom experience. Even though many people think of kids as having a natural desire to learn, they may feel that the typical classroom environment stifles rather than harnesses this instinct. One metaphor for this view is that kids are something like fires that need stoking (rather than, for example, plants that grow on their own with only a minimum of attention).

People are particularly likely to feel that it is difficult or even unnatural for most boys to do the things that are expected of students—sit still, be obedient, et cetera. Rather than seeing this is a moral (or any other kind of) failing on boys' parts, this view frames the problem as one of fit between boys and a particular environment. There is significant support for the idea that school, particularly in the earliest grades, is a better fit for girls than for boys.

> Boys ... like 11 to 12 ... they're just all over the place with their testosterone ... if they were ... doing the physical stuff [for example, chopping wood] ... they could blow off some of this excess energy they have, so they could focus better on their education. It's harder for young boys to concentrate and sit still and work on tasks and be more physically passive, and cooperate and collaborate and do group things.

This view would suggest solutions very different from the ones following from the "Defective Boy" model—in this case, the solutions would be more about changing environments so that they offer boys more of what they need, which could include, for example, opportunities for physical movement. Note that this is not a suggestion that schools abdicate their role in helping instill responsibility and self-control in children. Rather, it is a more positive framing of boys which offers the possibility of working with their strengths rather than against them. The notion of "working with the grain" may be a fertile angle to explore.

An interesting question raised by the study is whether the set of standards that make up "school morality" is more like "feminine" than "masculine" morality—for instance do expectations in school (and definitions of who is "good") fit better with feminine than masculine behavioral tendencies? While some aspects of school— particularly, competition in later grades—seem to play to stereotypical masculine strengths, many others are easier for girls to adapt to, with the result that girls are judged as "better," not only in the sense of competence, but often in the sense that they are better people.

> [O]n the whole most girls can tolerate the imprisonment of the old fashioned classroom that boys at a young age have trouble tolerating.

Well I think this open classroom worked very well. Because there's a lot of movement, and there's a lot of physical freedom. You're not free to disrupt, but you're free to move, you're free to make reasonable noise. You're free to move about.... But a kid like that in another classroom is labeled disruptive, and tied up and put away.

People are sometimes conflicted on the issue of gender equity in the classroom because of a clash between "public discourse" and "lived experience" on this issue. It is not uncommon for people to express conflicting views, sometimes only moments apart, on the difference between boys' and girls' experiences in the classroom. Consider this statement, in which the subject seems to contradict himself within the space of a single sentence:

I think [girls] identify with the teachers more, and the teachers are paying more attention to the girl students, which is great, because you know, you hear about girls being... left out and left behind, which I think is a shame because they're ALL our kids.

On the one hand the subject is saying that teachers pay more attention to girls, and on the other hand that girls are being "left out." Are girls being neglected or not? This person, like many people, seems simultaneously to hold two conflicting views.

In previous research, Cultural Logic has often identified a pattern which we refer to as "toggling"—individuals hold conflicting models and move back and forth between them at different moments. Typically, this toggling reflects two different sources of input—specifically, ideas that come from various forms of public discourse, such as the media, versus impressions based on the individual's own lived experience. For instance, Americans tend to toggle between the view that teenagers are reckless, selfish, and alien (a stereotype propagated in public discourse) and a view more based on their personal experience with teenagers, in which teens are more like less experienced, less confident versions of themselves (Aubrun and Grady, 2000).

In the case of boys and girls in school, some people appear to have one model based on certain strains in public discourse (what "you hear about" in the media, for instance) and another based on more direct knowledge, such as their own children's experience. Various forms of the Disadvantaged Girls or Defective Boys model have gained currency in public discourse, even though individuals may not have discerned the pattern in their own encounters with classrooms. As a result, people may experience something like "double-think," simultaneously believing that girls are discriminated against (public discourse), and that teachers and the school setting favor girls in a number of ways (personal experience).

Note that this is not to say that one view or the other is right. We have already argued that the Defective Boys model may lead to unfortunate consequences, but the point here is that the Schott Foundation has an opportunity to have an effect on public discourse, and therefore to help resolve a contradiction in people's minds.

Interestingly, as we have noted, experts have a greater tendency to think in terms of the Disadvantaged Girls model than lay people. There may be more than one explanation for this discrepancy: (1) Experts are aware of facts about education that the public doesn't know about; and/or (2) Experts are more conversant with the theories and the models that are played out in public (including expert) discourse, and are used to dealing with these models explicitly in their professional lives.

There are two conflicting ideas of the ideal classroom: one emphasizing individual students and teachers and the other emphasizing the "culture" of the classroom. The American model of school includes two alternative ways of thinking about the shape that the ideal classroom would take—the opposition between what could be called the "Teacher-Student" perspective and the "Classroom Culture" perspective. These aren't so much distinct opinions as distinct ways of understanding aspects of the classroom experience.

The first of these models—which is closer to "top of mind" understanding—emphasizes better relationships between the teacher and the individual students. Teachers who are as aware as possible of the strengths, weaknesses, backgrounds, and needs of each student could, in principle, create educational plans that would be ideally suited to their students. The term "individualized education" is an expression of this version of the ideal, and most people find it quite appealing (if impractical—see our discussion below). Nearly everyone is ready to say, when asked, that a teacher should try as hard as possible to take the needs of the individual child into account in creating an instructional plan. This view springs from an understanding of education as primarily being about communication of knowledge from teacher to student. A simple diagram of the key causal relationship in the classroom would look like this:

Teacher-Student Experience

The other model of ideal classrooms—more implicit, but very common nonetheless—emphasizes the culture of the classroom: the organic network of relationships and interactions among teacher and students, and especially among the students themselves. When people think about the culture of the classroom they are thinking about how it feels to be in class from day to day—for instance, is the atmosphere studious, energetic, passive, fearful? Teachers facilitate the classroom culture, and so their skills (as well as their personalities and attitudes) are still quite important, but this understanding of the class highlights a different set of skills.

The emphasis here is less on their ability to explain concepts, for instance, than on their ability to create an atmosphere of comfort and respect.

Teacher—Classroom Culture—Student Experience

The Teacher-Student perspective is more explicit, a view to which more people are likely to assent when questioned, probably in part because of the strength of individualism in American culture in general; students are thought of as unique individuals with individual purposes and trajectories in the classroom. But most people do directly or indirectly comment on the importance of the shared reality of the classroom, and focusing on this aspect of school has substantial advantages for funders. For one, it has the potential to liberate teachers from the daunting challenge of finding the ideal way of reaching each student; if they create the right atmosphere, there are benefits for everyone. This approach addresses the common concern that "individualized education" is more of an ideal than a practical reality.

There is some overlap between the individualized-instruction/classroom culture opposition and the very familiar opposition between a "regimented" classroom, where students sit quietly and absorb information dispensed by teachers, and a "free" classroom, where students are more active, both in the sense of physical movement and with respect to their role in their own education. The former model is more closely related to Lakoff's "strict father" model—teachers instruct, push, discipline, and punish their students. The latter is more closely related to Lakoff's (2002) "nurturant parent" model—teachers encourage their students, give them freedom to explore their own ways of learning, and take some cues from them in setting the agenda for class. People certainly differ with regard to which of these two models they regard as more ideal. On the other hand, some are inconsistent about which is better, and the regimented model seems to be more of a prototype in most people's minds. As a consequence, the regimented classroom represents an image which they fall back on as a default from time to time, even if they strongly support the idea of active classrooms (for instance there is "toggling" on this issue).

> [The] ideal classroom is one where the children feel that it's their classroom, and that they're in control of the learning. That the teacher is a resource for them. That there is no teacher's desk at the front of the room or anywhere, really. The classroom is a cooperative classroom, and the learning is organized to build that learning community . . . individual attention to individual students is absolutely important, but that everybody's individual attention can be everybody's learning opportunity. So, we could, you know, share in that way.

Key elements of social/emotional experience are seen as playing a critical role in academic success. While laypeople sometimes fall back on a simplistic separation between academic performance and other aspects of a child's personal experience in school, they more often demonstrate an awareness of the strong causal connection between these two domains. These key concepts, associated with the social realm, show up in discussions of all aspects of gender equity.

Expectations is an issue that applies to the classroom as well as to other areas of life. Just as parents should expect the most from their children, teachers are seen as needing to be available to lend additional support to kids having trouble reaching their full potential. Attention is a social factor that has obvious and critical importance in equity issues: The lack of attention can undermine the fairest, highest expectations. In an ideal learning environment the teacher is not spread too thin. She or he provides support and helps to motivate so that all students can learn. Teacher attention unevenly distributed effectively undercuts a favorable student/teacher ratio for those fortunate enough to have one.

People see children's comfort level as a critical component of their success in school. School is seen as an intimidating place on a variety of levels. Children are required to function publicly under various adult demands, in an environment which may be radically unlike their home life. One important purpose of schooling is socialization. (Many people mention this as the principal argument against home schooling, for example. This presents an interesting contradiction to their preference for teaching that comes as close as possible to a one-on-one relationship.) School is seen as a good social environment—*the* place—for children to learn how to work together, which is the argument that counters the frequent observation that boys and girls are distracting to each other.

Furthermore, mutual respect is seen as a critical element in a comfortable classroom environment.

The issue of gender equity plays out differently at different age/grade levels. When people think of "school," they tend to think first of elementary school—perhaps prototypically of the fourth and fifth grades. It is important to keep in mind that some aspects of the cultural model of school do not apply equally well to high school as to elementary school. In particular, the idea that school is a kind of "bubble," existing within society but not quite of it (see above) is less prominent in the case of high school. High school students are older—more like the adults who make up the larger society. In some ways, a high school is more like a workplace, where one expects issues of gender inequities to crop up. In addition, as students become teenagers, issues of sexual harassment—as well as issues of ethnic and gender identity and orientation—become more salient, all of which contributes to a lessening of the "bubble" model of school.

Most people believe that there are important differences in how teachers should communicate with different individual children, but not different groups (for example, ethnicities). There is near unanimity that teachers should tailor communication to individual kids. Many people go beyond that, believing that a teacher should get to know each child: their strengths, weaknesses, family background. Most people are unaware of, or wary of, considering possible differences in communication style among different ethnic, racial, or gender groups. Yet a minority of people feels quite strongly that for some children the gap between the school atmosphere and their home life is a chasm. Some evidence suggests that the lack of emphasis on tailoring the communication process to different groups is due to the fact that it contradicts the powerful cultural model of the classroom where a common (American) identity is forged. The ideal classroom culture in which the teacher knows the students and the students develop in a social environment of respect and cooperation, is seen as reducing differences between children, including differences based on race, class, and family circumstances.

People have a consumer model of education: it's mostly about getting the best quality and best advantage for their own kids. Beneath their stated reasons for supporting public schools—people are quick to share such mantra-like statements as "kids are our future" or "it's the fundamental responsibility of a society" to educate its young, parents tend to have a *consumer model* of education. Their concern is first and foremost with getting the best quality for their own kids. More than anything else, they want their kids to *learn*. For example, parents feel cheated when kids are passed before they've learned the requisite amount. Ideally their kids will be engaged and excited by school; but the bottom line is that the school system should ensure that their children will have all the advantages that other kids have ("everyone should have pretty much equal chance"). Any communications message must take account of the consumer perspective on education—quality. Education is about individual achievement, benefit, and advancement. Self-betterment through education and hard work is one of the lynchpins of American individualism. Its outcome is expected to be personal gain—epitomized by upward social mobility.

Interestingly, many people articulate a more communitarian vision of education, with a parental role more active than mere passive consumer. Yet when pressed, most interviewees see "parental involvement" as nothing more than monitoring their children's homework completion. The more fundamental (if less directly expressed) consumer view is evident in the overriding concern with quality. Some education consumers actively "shop around" for the best schools and even relocate to give their kids access to the best districts.

Specific skills—especially literacy—are important to people. Parents think, Is my kid learning to read? before they think about whether there is gender equity. It follows that tying gender equity to "quality" issues should be effective. For example, for the people who believe that teacher attention is unfairly distributed by gender (with potential neglect of girls versus boys, or vice versa), this link should be drawn out.

Conclusion

This chapter has explored the public's "mental map" of the relationship between gender and the school experience, and described ways in which perceptions of the issues go beyond the simple thesis that one sex or the other is specifically disadvantaged by current conditions in the classroom. Most people are aware of and accept the fact that serious inequities (based on gender, race, and class) exist in American society, they draw a strong distinction between the larger world and the classroom. Indeed, most people are either barely aware of the role of gender in defining students' experience, or feel that the problems are more symmetrical: (1) There is a set of stereotypical gender-based expectations that constrains both boys and girls and diminishes their experience; and (2) By its current nature, schools often provide a poor fit for boys and girls, in different and specific ways (and more so for boys). A minority view that girls in particular are shortchanged in the classroom seems to be tied to some unwarranted and potentially damaging framings of boys.

In the brief discussions below we summarize differences in thinking across two salient fault lines: lay people versus experts (for instance individuals with a close involvement in decision making about education) and conservatives versus liberals (as we define these below).

Experts versus Lay People

Naturally, experts in education are more knowledgeable than the public at large about conditions in the classroom. As a result, they are more aware than the rest of the population of differences between boys' and girls' experiences in school. On the positive side, this means that they tend to be ready to think about solutions, and to have already given the issues a fair amount of thought. In some ways, though, they may prove to be a tougher audience for the Schott Foundations' ultimate campaign, whatever its content might be. This is because experts are more likely to be comfortable with particular theories already—for example, the commitment to the Disadvantaged Girls/Defective Boys model which we encountered among some professionals appears to correspond with particular trends in public discourse on the issues—and more likely to adopt an authoritative stance toward the issues.

Additional points to note about experts are that (1) they are more likely to be open to the possibility of tailoring educational practice to particular groups (for example, ethnic groups) of students, and therefore more open to tying gender issues to identity issues more broadly; and (2) that the consumer model of education ("What quality of education is my kid getting?") is less appealing to them than it is to others in the community. This second point means that experts should, overall, be more open than the public as a whole to messages that seem less directly tied to bottom-line issues like test scores.

Conservatives versus Liberals

Two caveats are required here. The first is that the population we discussed these issues with does not represent an accurate cross-section of adults in Boston, Cambridge, and Western Massachusetts—adults who are concerned enough to commit time to talking about education are, to an extent, a self-selecting group. And in fact, self-described conservatives are not as well represented in our sample as self-described liberals. On the other hand, and as predicted by cultural model analysis, the division between political conservatives and liberals is less significant than a division between "cultural" conservatives and liberals. That is to say, people's implicit understandings of daily life do not always correspond with their explicit theories of how society should be organized (for example, as reflected in how they vote).

The most general conclusion from a comparison between conservatives and liberals is that the distinctions do not fall out in the simple ways one might expect. First, the idea of traditional sex roles seems to play no role in most people's current thinking—the idea that conservatives, for example, are less interested in girls' education than in boys' is probably false, outside of certain restricted populations (some religious fundamentalists, for example). The individualistic consumer model favored by conservatives is clearly compatible with some forms of gender equity—conservatives have daughters and think of them as consumers who deserve the best.

But there are differences between the groups. For example, cultural conservatives favor regimented classrooms, while progressives are more likely to favor free classrooms. More directly relevant to gender issues, cultural conservatives seem less likely to support the idea that each student's cultural background must be taken into account in the classroom, instead favoring the idea that schools, among all our institutions, should most exemplify the American notion of the "melting pot." As a consequence, conservatives are probably not predisposed to respond positively to a campaign based on tying gender to other identity issues.

The good news for foundations is that these starting differences will not necessarily lead to schisms on particular issues. To take just one example, conservatives' reliance on the "melting pot" assumption can lead them to a "culture of the classroom" view, which jibes in various ways with many liberals' belief in an "organic" model of the classroom. In short, both conservatives and liberals are potentially open to a number of messages, depending on how the issues are framed.

Note

This chapter was originally published in 2000 as "Gender Equity in Schools." It is collaboration between the Frameworks Institute and the Schott Foundation.

References

Aubrun, A., and Grady, J. (2000). *How Americans understand teens: Findings from cognitive interviews.* Washington, DC: Frameworks Institute.

Lakoff, George. (2002). *Moral politics: How liberals and conservatives think.* Chicago: University of Chicago Press.

Chapter Two

Gender Matters: Funding Effective Programs for Women and Girls

Molly Mead

In 1992, the executive director of the National Council for Research on Women, Mary Ellen Capek, stated publicly that having either the word "women" or "girls" in the name of an organization seeking foundation funding was the "kiss of death." In short, it was a sure bet that such an organization *would not* be funded. Her statement reflected anecdotal evidence from foundation program officers, who had seen too many funding applications from women's organizations be turned down because of their gender focus. Additionally, her pronouncement was buttressed by national and regional research studies, which documented how the philanthropic community responded to women and girls—neither supporting organizations for them proportionate to their population or their growing needs.

Even as Capek offered her grim but realistic assessment of the funding prospects for women and girls in 1992, a major national effort was underway to increase foundation funding for these programs. Both national and regional organizations were engaged in an interlocking set of activities designed to educate funders about women and girls and to provide them with solid evidence of the ineffectiveness and inequity of their own grantmaking practices. The assumption was that such evidence would be a powerful force for change.

What have been the results of these efforts? What (if anything) has changed since 1992 in funding for women and girls' organizations? What is the best evidence today for why foundations should consider gender in their grantmaking? In addition, what strategies should advocates pursue to influence the ways foundations incorporate gender considerations in their grantmaking priorities? In this chapter, I answer the above questions by analyzing the most recent thinking about these issues. To do so, I have divided this chapter into three sections: The Current Status of Foundation Funding for Women and

Girls, The Reasons to Incorporate Gender Analysis in Grantmaking, and How to Change Foundations' Approach to Gender in Their Grantmaking.

The Current Status of Foundation Funding for Women and Girls

National and regional data on the percentage of foundation funding for women and girls shows a consistent pattern on both the national and regional level: foundation funding for women and girls hovers in the range of 3 percent to just over 6 percent of total grantmaking. Research on grantmakers' attitudes about funding women and girls shows the reasons for this funding pattern. The bottom line is that funders have a strong preference for funding so-called universal (or coeducational) programs, and little awareness of the need to consider gender when setting grantmaking priorities or allocating funds to grantees.

The Reasons to Incorporate Gender Analysis in Grantmaking

There are substantial reasons why grantmakers should systematically include an analysis of gender considerations in their grantmaking. Although gender is a socially constructed category it serves to both proscribe and constrain the life experiences, opportunities, access to resources, and, finally, power balance between women and men. Because of the power of gender to shape the lives of women and men, both individually and collectively, women are often differentially or disproportionately affected by many of the public problems that foundations profess to care about. In addition, women often benefit *less* from public policies that are developed (ostensibly) to benefit everyone *equally*. This can be a drag on democratic ideals. For all these reasons, it is imperative that funders incorporate gender analysis in their grantmaking—if they want it to be both effective and equitable.

A powerful case example of what can happen when grantmaking ignores the issue of gender is provided by coeducational youth development programs for urban teenagers. I studied 25 such programs to see whether girls are being well served in these "universal" programs. What I learned is that girls and boys need some different programs and approaches because of differences in life experience and gender norms. Unfortunately, this is not recognized by these so-called universal programs. While gender differences in girls and boys are neither innate nor immutable, a program that aims to be fully effective for girls must incorporate gender considerations in its program design and operation. This is best done by exploring the social construction of gender and inviting young women and men to challenge gender norms, examine gender privilege, and create an even balance of power between girls and boys. Although my research focused on a narrow population (teenagers living in urban communities) and on only one type of program approach (youth development), my conclusions have potential applicability to other ages and other program approaches. To be effective for women and girls, programs need to take gender into account. To take one step back then, funders of these programs also need to take gender into account.

However, gender cannot be considered in isolation from *other* socially constructed categories that also constrain and determine access and opportunity. Thus, it will not suffice for grantmakers to incorporate a single-gender lens in their grantmaking in order to be fully responsive to women's social location. A uniform approach is likely to lead to ineffective grantmaking as well as legitimate resistance. Recent scholarship done on race, class, gender, and sexuality has helped us understand that there is *no one* gendered existence. The ways in which the category of gender shapes lives and constrains opportunities is historically and geographically contextual. In addition, women and men are not *simply* gendered beings. Individually and collectively, our race, our socioeconomic class, and our sexuality shape all of us, to name other important categories of social-group membership. What this means in the United States, for example, is that the experiences and opportunities of white, upper-middle-class women are often quite different from white working-class women, and different yet again from women of color.

Ironically, grantmakers who understand these differences have proven to be a major source of resistance to adopting a gender lens. These grantmakers, when presented with arguments that they should adopt a gender lens in their grantmaking, have correctly responded that issues of race and class operate in powerful ways and must also be considered in their work. A way out of this dilemma is to think about gender lenses, or, even better, gender analysis—versus a single-gender lens—to acknowledge the simultaneous existence of several socially constructed group memberships. This could result in better grantmaking and a more powerful consensus about the need to incorporate gender in grantmaking. Although there are some daunting practical challenges associated with adopting gender analysis—namely that it requires well-trained program officers—it is imperative if the desired outcome is the most effective grantmaking possible.

I conclude with a summary of the literature on universal and targeted funding and place this discussion in the larger public policy debate about universal and targeted policy approaches. Defining the concepts universal and targeted as a binary can be misleading because oppositional thinking forces a set of either/or choices that is neither realistic nor optimal. Indeed, it is both possible and desirable to combine the best elements of universal funding and targeted funding in a way that results in grantmaking that is efficient, effective, and popular.

Unfortunately, many of the efforts to increase foundation funding for women and girls have had the *unintended* consequence of *reinforcing* the notion that universal and targeted funding are contradictory choices—one of which works to women's advantage and one of which does not. Because most of our work had the effect of focusing attention on foundation grantmaking to programs specifically designed for women and girls, we helped create the conclusion that *the only way* to advance women was to engage in targeted grantmaking. I now believe that targeted grants are *one viable approach* to advancing the lives of women and girls, but that universal grantmaking can work just as well if it is done as part of a *"gender and"* analysis.

What goal should foundations strive for? Based on current research, they should fund a mix of effective programs. Just as there is evidence about the need for gender-specific programs for women and girls, there is also evidence backing up reasons to have universal programs that pay appropriate attention to gender. But the current funding percentages do not, on the face of it, appear to be justifiable. (The funding breakdown in Boston, for example is 92 percent coeducational, 6 percent all-female, 2 percent all-male). The funding for all-female programs seems especially low, given current evidence about the ways in which universal institutions underserve women.

How to Change Foundations' Approach to Gender in Their Grantmaking

The evidence presented in this chapter leads to one conclusion: *foundations must incorporate gender analysis in their grantmaking if they want their work to be effective.* The lessons learned from international grantmaking show that this could be accomplished, yet efforts thus far to change grantmaking practices in the United States have been ineffective. The bottom line is that the percentage of foundation funding for programs for women and girls has changed very little in the past ten years. We must ask once again, What can be done to ensure that gender analysis is institutionalized in the foundation world? There will not be a single answer to this question. As with all attempts to change well-established practices of major institutions, we will need to develop a multiplicity of approaches. We may not even agree what those approaches should be.

In this chapter, I also explore what could be done to change grantmaking practices of foundations. To start this dialogue, I survey the relatively meager literature on how to change foundation behavior. What I conclude is that foundation grantmaking behavior needs to be understood on several levels simultaneously. Foundations are both rational *and* irrational in their decision making: they are influenced not only by carefully presented research evidence but also by internal and external pressures. My analysis, therefore, argues for a multipronged change strategy, offering evidence of the continuing relevance of gender to grantmaking; education of domestic funders on the institutionalization of gender analysis in international grantmaking; and a carefully orchestrated internal and external campaign for change.

The Current Status of Foundation Funding for Women and Girls

By now many of us know that there is a bias in organized philanthropy against funding programs that are designed specifically for women and girls. In the aggregate, foundations give a very small percentage (usually 6 percent or less) of their funding to programs specifically designed to work with women and girls. This section explores the reasons behind this funding pattern.

Universal Programs

Program officers and executive directors of foundations have a strong preference for funding *universal* programs (programs that serve males and females) over gender-specific programs, and they do not include gender as a major determinant when they set funding priorities. In general, most foundation staffs are convinced that their current approach (developing funding priorities without considering gender and funding universal, nongender-specific programs) is adequate *and even preferable* for meeting the needs of women and girls.

Foundation staff members *do* express interest in the need to alleviate the most pressing problems of women and girls and to create programs that develop their strengths and pave the way to full and active membership in a civil society. Perhaps most important, they are *convinced* that their current funding priorities *do just that*—a viewpoint very much disputed herein.

At first glance, it might seem that the patterns of foundation funding are relatively irrelevant to nonprofit programs, since foundations typically provide a relatively small percentage of an organization's funding. In the aggregate, foundation funding provides approximately 12 percent of the budgets of programs for women and girls. Yet foundation funding comprises a much larger percentage of the budgets of the smaller and newer organizations. In Boston, for example, foundation funding accounts for 40 percent of the funding of organizations with budgets under $125,000; 69 percent of the funding of organizations with budgets under $37,000; and 51 percent of the funding of the newest organizations: those in existence for less than three years (Mead, 1994). What these figures show is that foundations play a particularly significant role in supporting innovation and in funding organizations that might have trouble getting funds elsewhere.

National and Regional Research on Funding For Women and Girls

According to national figures compiled by the Foundation Center for grants awarded in 1999, only 6.4 percent of all foundation dollars were designated for programs that specifically benefit women and girls (Lawrence, Gluck, and Ganguly, 2001). Researchers have been tracking this percentage since the 1970s, when two reports focused the attention of the philanthropic world on the relatively small proportion of foundation dollars that were *intentionally directed* to women and girls. In 1975, Mary Jane Tully reported in *Foundation News* that less than .5 percent of all foundation funding went to programs specifically for women and girls. In 1979, the Ford Foundation issued a report documenting that (in 1976) only .6 percent of the more than $2 billion in foundation grants were specifically designated to benefit women and girls (Ford Foundation, 1979). Since 1981, the Foundation Center has tracked grants to women and girls. In order to be classified as a grant to women and girls, a grant must meet one of the following four requirements:

- Women and girls make up a substantial majority of either the members or clients of the agency or program.

- The agency or program is intended to increase participation by, or extend services to, women or girls.
- The agency or program addresses an issue or discipline as it affects women or girls.
- The agency or program addresses an issue or discipline whose impact is primarily upon women and girls.

Although grants classified as being made specifically to women and girls do not reflect the total amount of funding that reaches women and girls, they do reflect grants *purposefully targeted to them*. In that respect, the dollars of giving reported by the Foundation Center reflect grantmakers' *intentional and specific funding* for women and girls. One can conclude, therefore, that these percentages are an important indication of *how philanthropists view the need for programs designed for women and girls*.

The Foundation Center's data has several limitations that are worth mentioning. First, many women's funds argue that an important component is absent from the classification criteria—that an organization is controlled by women (as represented by staff leadership and board membership). Second, the data excludes grants under $10,000. While no one argues that this exclusion significantly skews the total, it does render invisible the important work of many women's funds, which typically make grants under $10,000. Because there is another organization (The Women's Funding Network) that does track and report all grantmaking (without exclusions) by women's funds, it would be ideal if the Foundation Center could incorporate this data as well into their reports. Although the percentage of foundation dollars to programs for women and girls from 1975 to 2003 increased in this 24-year period from .5 percent to 6.4 percent, even the latest percentage continues to reflect a minuscule portion of philanthropic dollars (Foundation Center's Statistical Information Service, 2005). Based on the foregoing, it is clear that funding programs specifically for women and girls is *not* a priority to the majority of foundations.

Where Funding Goes

The research on funding patterns in Greater Boston may serve to explain *why*. This study of foundation allocations to programs for women and girls asked two additional questions, which were natural outgrowths from the earlier studies: (1) Where does the 95 percent of foundation allocations go? and (2) Why are the allocation patterns the way they are? (Mead, 1994).

In the Greater Boston area, 92 percent of foundation dollars goes to universal programs; 6 percent goes to programs for women and girls; and 2 percent goes to programs for men and boys. Overall, funders overwhelmingly give their money to programs that do not specify a gender focus.[1] This finding raises a central question: Are universal programs effective for women and girls? It turns out that most foundations simply *assume* the answer is yes. Unfortunately, they do not have the data to support their assumption.

Funders Assumptions about Where Funding Should Go

Only 20 percent of the funders surveyed in the Greater Boston Study think that funders have historically underfunded women and girls' programs. Most (61 percent) believe that their current funding strategies appropriately benefit women and girls. Funders in Greater Boston cite a variety of reasons for preferring universal programs. In focus groups and individual interviews they revealed the reasons why they prefer universal over gender-specific programs.

Their responses can be grouped into five categories, each of which reveals serious flaws in thinking.

Efficiency. Given limited grantmaking dollars, the best investment is to give the money to nongender-specific programs. Some grantmakers conclude that programs that serve both males and females are a better investment than programs that work with only one gender. Two-fifths of the Greater Boston Study respondents specifically oppose a gender focus because, as one stated, "[W]e have limited resources and want to reach the broadest audience with our funding." Another program officer stated his view in very practical terms: "If we get two proposals for similar programs, one to work with males and females and one to work with just females, we would probably fund the male and female program."

Democracy. Targeted programs promote exclusivity. Most foundation and corporate-giving program staff are interested in a democratic society. Some grantmakers view gender-specific programs as promoting exclusivity and creating separate worlds. To these foundation professionals, the idea of a single-sex program is counterintuitive. They posit that the advancement of girls and women is likely to occur in programs that treat male and female constituents equally and therefore identically. These funders conclude that a gender-specific program runs counter to the gender equity they imagine for society. Others conclude that a female-focused program operates too far out of the mainstream of society. As one executive director stated in an interview, "organizations that are run by women, that exclude men, and whose purpose is to advocate for women in a political context are too exclusionary and go too far."

Efficacy. Targeting hurts women and girls. Some philanthropic staff members think it is counterproductive to the interests of women and girls to create programs that work exclusively with them. They point out that the world consists of men and women. They question whether females can ever gain equity with males if they don't learn to live with and confront the gender inequities that occur. As one program officer stated, "In an ideal world, I hope that there will be no more boys' clubs and that we will have only boys' and girls' organizations. There are girls' and women's issues, but they have to be addressed systemically. I don't think that an after-school program for girls can really change self-esteem, especially when girls live every day in classrooms with teachers who favor boys."

Relevance. Gender is an irrelevant category for targeting. Some grantmakers believe that gender is not relevant when they make strategic decisions about their grantmaking. As one foundation executive director put it, "In the Boston grantmaking community women's funding is a non-issue. The lines are drawn more on social, cultural, and economic dimensions. Gender is not a critical criteria." This category of grantmakers concludes that there *may* be reasons why it is legitimate to fund programs that serve a specifically defined constituency and that such programs do not violate previously defined concerns about efficiency or democracy. Yet they do not conclude that gender constitutes an appropriately defined constituency. They argue that there *are* reasons to fund programs that serve a racial, ethnic, or economic constituency, but there are *not* good reasons to fund programs that serve a constituency defined by gender. As one program officer put it, "It is easier to sell ethnic than gender advocacy. It is harder to be convinced of the need for women to organize as a separate group compared with those who have such barriers as language or recent immigration and who lack the resources most women have."

We don't fund women. Our organization's philosophy and mission does not include a focus on women. Some funders said that their entire grantmaking strategy obviates consideration of gender as a category—either because they don't fund specific population groups or because they don't identify women or girls as a category relevant to their mission, their funding guidelines, or their founder's intention. Rather, in some foundations, grantmaking criteria point in other directions. As an executive director of one foundation said, "We want to fund communities, not specific populations." Another executive director noted, "Our primary focus is to fund low-income people as they reside in the neighborhood setting. There are times when there is a challenge between the fit with a proposal that is just framed as addressing women's and girls' needs, and our neighborhood framework."

Examining Funders' Assumptions about Gender

At first, it may seem that these categories of responses merely illustrate the existence of a set of logical criteria to determine whether a foundation will consider gender in funding decisions. Further analysis, however, uncovers major limitations in their thought processes, revealing that their funding criteria is based (in part) on a lack of understanding of both gender as a relevant category in grantmaking and the role that programs for women and girls plays in their lives.

Reason #1
Efficiency: You cannot be efficient if you are not effective. While funders are justifiably concerned with efficiency in their grantmaking, a grantee agency is not automatically efficient because its constituents are male and female. A universal agency must work reasonably well for its entire constituency in order to merit funding. However, the Greater Boston Study revealed that funders usually do not ask universal programs to report even the most general evaluation data by gender. As a result, funders do not know how well these programs are serving the needs of women and girls.

Reason #2
Democracy: Single-sex programs can produce democratic outcomes. Funders concerned about democracy must examine their assumption that single-sex programs are automatically antidemocratic. Research, in fact, has shown that programs designed exclusively for a "minority" group often develops strengths in that group that allow them to function even more effectively in the "majority" world. Studies comparing coeducational with single-sex schools, for example, support the view that single-sex schools produce better outcomes for girls in both cognitive and social measures (Riordan and Lloyd, 1990; Sadker and Sadker, 1994). Women who attended single-sex colleges have higher educational and occupational achievement, higher self-esteem

and more supportive views of equal sex roles than women who attended coeducational colleges (Riordan and Lloyd, 1990).

Reason #3

Efficacy: Girls have plenty of opportunities to learn how to live on an equal footing with boys. There is no need to recreate sexism. The funders whose attitudes are captured in the efficacy argument described above make a revealing assumption about program effectiveness—that women and girls, by being forced to encounter sexism, learn necessary survival skills when they participate in universal programs, skills that are not developed in single-sex programs. This is both a harsh view of gender relations and also assumes that women and girls should be "toughened up" to deal with sexism rather than believing sexism can or should be challenged. It is hard to imagine, if funders reflected carefully on their own reasoning, that they would continue to suggest that funding sexist programs helps prepare girls for a sexist world. Yet that is what this reasoning amounts to.

Reason #4

Relevance: Women are found in all the other "categories" funders consider relevant and their needs are often different from the men in those categories. The assumptions embedded in the argument that "targeting" by race, ethnicity, or socioeconomic status is appropriate, whereas gender is not, are the most troubling because they reveal that funders equate women with white, upper-middle-class women, and they equate racial constituencies with men. This assumption makes invisible the many women of color and low-income women who are served by (and also run) programs for women and girls. It keeps in place a double bind for women of color: that "women" are white and "people of color" are male. In fact, the boards, staff, and constituents of most programs for women and girls are racially diverse. Again in contradiction to some funders' assumptions, these are not organizations exclusively for or run by white women. In 24 percent of the organizations surveyed in the Greater Boston Study, people of color account for the majority of staff. In 33 percent of the organizations, the majority of clients are people or color. In 18 percent the majority of board members are people of color; and in 12 percent the director is a woman of color. Overall, the organizations in this study report that 47 percent of their clients are white. Roughly a quarter is African American, 13 percent are Latina, 5 percent are Asian, and less than 1 percent is Native American Indian. The racial diversity of these organizations is especially significant in light of the region's overall demographics: 83 percent of the girls and 90 percent of the women who live in Greater Boston are white.

Reason #5

Make Women Visible: You will have more success accomplishing the mission of your foundation. The funders who say that gender simply does not fit anywhere into their grantmaking (because the focus of their foundation is in entirely different areas), often fail to see women at all.

Women and girls, after all, live in neighborhoods and often play important roles in the conditions of those neighborhoods, and more women than men live in poverty. So, a foundation concerned with community development or with low-income people might naturally also be interested in the role of gender in these issues. However, the responses above indicate that they do not make this connection. In many funders' minds gender is a discrete category which is treated as separate and distinct from categories like community or income level. This way of thinking about gender must be addressed if funders are to ever change their approach to funding women and girls. Otherwise women will be viewed as being in a (win-lose) competition with other groups not defined along gender dimensions, but which also have equity interests.

The Relevance of Gender to Foundation Grantmaking

Despite their many differences, foundations have one thing in common: they seek to fund organizations that are effective in accomplishing their goals and objectives. The primary reason, then, why foundations need to consider gender in their grantmaking is that many of the programs they fund simply are not as effective as they could be because they fail to incorporate gender considerations in their design and implementation. This section will discuss the need to consider gender in the design, implementation, and leadership of the programs that foundations support.

Unfortunately, there has been a trend in the last twenty years to de-emphasize gender in program design; ironically, this de-emphasis has arisen in part from efforts to *ensure* gender equity. Simply put, efforts to equalize women's status with men's often have been equated with the idea that a woman can do anything that a man can. This idea made sense as a strategy to ensure that women gain entrée into previously male domains and programs; but it also resulted in a blurring of the recognition that there continue to be significant disparities in access to resources and opportunities between men and women. It also had the unintended effect of continuing to hold men as the standard to which women should be compared.

At one time, the rationale for paying attention to gender in program design was the supposition that women and men were inherently different, and women, as the subordinate group, needed special programs to compensate for those differences. This rationale has (and rightly so) been challenged. The challenge, however, often resulted in the opposite conclusion—that women and men are just the same and therefore universal programs should work as well for a woman as for a man. In practice, this has often resulted in women's admission into (formerly all-male) programs that were designed to work well (only) for men. While research has indeed shown there are few inherent differences between women and men, there are *profound* differences in the socialization of gender that impact women's and men's lives.

The central case, then, for the relevance of gender to program effectiveness rests on the fact that men and women are socialized very differently, with different expectations about appropriate behavior and social roles, and that society holds men and women to different standards of behavior. This different socialization leads to different life experiences and opportunities that can result in quite profound gender disparities in access to tangible and intangible resources.

If these programs want to be fully effective, they must pay attention to these differences. Foundations can play an important role in this context. First, they can fund established programs that have a strong track record of working effectively for women. Second, they can motivate programs to incorporate gender concerns by including gender criteria in grantmaking decisions.

However, foundations (and programs) cannot pay attention to gender in isolation from other socially constructed categories such as race and class. As with gender, people's lives are profoundly shaped by racial, cultural and class contexts, which impact experiences and opportunities. Well-designed programs need to pay attention to gender considerations in conjunction with race and class considerations. Foundations need to do the same. Although it may seem obvious that race, class, and gender altogether shape individual and group experience, it is actually easy to lose sight of this fact. This is due in large part to the fact that the predominant image of a woman is white, professional, and upper middle class, despite the fact that most women in the United States are not white and upper middle class.

This misconception obscures the fact that gender, race, and class are inextricable; that one is not lived without the other. When we lose sight of the intersection of race, class, and gender we can fall into thinking that race and gender are opposite categories, only one of which can be considered at a time. This false dichotomy is evident in foundations that include racial criteria in their grantmaking but not gender criteria. This practice pits white women against people of color and excludes women of color altogether. The way out of this conundrum is to recognize and analyze the intersection of race and class with gender both in programs and in the funding of those programs.

This section concludes with a discussion of the relevance of gender to youth development programs. The evolution of youth programs provides a powerful, concrete example of the ideas presented in the remainder of this section. Previously, youth programs were primarily single-sex. This separation was grounded in the widely accepted (but false) belief that boys and girls were inherently and inevitably quite different from each other with totally different needs, strengths, and interests. When that belief began to be questioned, and when the disparity in numbers of youth programs for girls was identified, there was a strong push to open up boys programs to girls.

The new thinking about gender was that boys and girls were not so different from each other after all, and hence would do equally well in the same program. In practice, however, many girl-only programs went out of operation; boys programs became coed; and girls then had to fit into programs designed for boys. Finally, the new thinking about gender often failed to account for socialized gender differences that resulted in girls and boys having different life experiences. What the youth program example shows is that effective programs need to recognize and challenge the socialization of gender.

In the arena of international development there is strong and widely accepted evidence that development programs come closer to achieving their goals when women are considered in the design, implementation, and evaluation of development activities. In this arena, the current direction is to develop both universal and women-only programs that pay specific attention to the needs of women and that involve women in the design and delivery of the programs.

Gender Matters

The fact that women and men are socialized differently and held to different standards of behavior by society means that programs must incorporate gender in program design, implementation, and leadership to be fully effective. The following categories provide specific examples of the different ways gender "matters." The first five examples argue for thoughtfully designed programs that might be either coeducational or single-sex. The last three examples argue for the need for well-designed, gender-specific programs.

- *Disproportionate impact of specific public problems.* In some instances, because of their social roles, women and girls are *more* affected by a specific problem or issue than are men or boys. For example, it is well-documented that there are significantly more women than men living in poverty. This is primarily due to two factors: labor-market segregation and women's significantly greater role in raising children. Overall, women work in jobs that pay less (than men). Moreover, because they often work fewer paid hours due to childcare responsibilities, this results in more women [than men] being poor. Therefore, poverty and all its attendant issues have a greater impact on women than men. A program concerned with issues of poverty or works with poor people then needs to do a gender analysis to verify that it is effectively reaching the group most affected.

- *Differential impact.* Women and girls also can be impacted *differently* (than men and boys) by a problem. For example, the opportunistic infections which women typically manifest when they have full-blown AIDS are different from the opportunistic infections that men manifest. Before this difference was recognized, women with AIDS were excluded from receiving any of the benefits of programs which were designated for *all* people with AIDS, but designed according to a *male* definition of a person with AIDS. Although the variation in opportunistic infections stems from a biological difference between women and men, the male-centered definition of an AIDS diagnosis arose from the still widespread practice of making male health the norm for everyone. Spending less money on research on women and AIDS or basing a definition of AIDS only on men's diseases is a social issue, not a biological one. Programs designed specifically for women with AIDS were at the forefront in advocating for the needs of women with AIDS, pushing the Centers for Disease Control (CDC) to change its definition of AIDS, so that women could receive the federal and state services that accompany an AIDS diagnosis.
- The different social roles that women and girls occupy (for example, raising children) also argue for differences in program design. Labor-market segregation, resulting in lower wages for women compared to men, needs to be addressed in job-training programs for women. Additionally, if they expect women to participate in such programs as substance abuse treatment, programs must take into account the childcare responsibilities that many women hold.
- The different socialization of women and girls should also result in the need for appropriately designed programs. In general, women and girls are socialized to play a caretaking role and to be subservient to men and boys. Thus, when women and girls are in

mixed-gender groups, they may talk less, may venture fewer opinions, and may be reluctant to engage in verbal conflicts. All of these behaviors will influence the success of a program. In a coed leadership program, for example, girls may be less willing to engage in behaviors that are traditionally associated with leadership. Such a program will need to design specific approaches to counter this socialization if it expects to work as well for girls as for boys.
- Gender socialization can also result in different opportunities available to women and girls. Despite little difference in inherent mathematical ability, for example, girls take fewer advanced math and science courses than boys. This significantly constrains future career choices.

Sometimes Women-Only Programs Make the Most Sense

The examples provided above all make explicit the ways in which gender matters very much in program design and implementation. They each refute the contention of those foundation professionals who believe that gender is not relevant to their grantmaking. They don't, however, resolve the debate about whether women and girls are better served in universal or single-sex programs. If a universal program acknowledges the multiple ways gender socialization affects its constituents, it can develop an effective program for women and men. However, single-sex programs can offer some advantages that universal programs, no matter how well designed, cannot. These advantages are summarized below:

Avoiding feeling like the "other." There is considerable evidence that, in many ways, men and boys are viewed as "normal" and women and girls are viewed as "other." For example, an often repeated study by Broverman et al. (1972) shows that men and women rate the qualities we generally associate with the *male* role as being identical with the qualities of a psychologically *healthy* person. The qualities normally associated with the *female* role were virtually identical with the qualities of a psychologically *unhealthy* person. All too often, girls and women are compared to boys and men and found wanting. In a program in which every participant is female, the issue of being "other" recedes into the background, and participants can concentrate more on being themselves, working to develop their unique capacities and identities. It is ironic yet true that in a single-sex program, the issue of gender can become a non-issue and other concerns can then take precedence.

Feeling safe and being safe. Another reason to provide programs for women and girls is that they can feel more comfortable and safer in an all-female environment. The issue of safety is hardly trivial. Women and girls are less safe in their homes than they are in public places and they are used to feeling unsafe in environments that men and boys perceive as safe. For too many women and girls, unfortunately, men and boys are the greatest threat to their safety, and too often this includes men and boys enrolled in universal programs. Women who are homeless, for example, often report that a homeless shelter is the choice of last resort because of the frequent sexual violence occurring there. Girls in coed-youth programs report high levels of sexual harassment by the boys in those programs. Sometimes the location itself of a universal program is so unsafe that it presents an overwhelming barrier for women and girls.

Women and girls are in charge. Typically, women and girls run single-sex programs for women and girls. In my research on nonprofit programs in Greater Boston, I found that every program for women and girls is run by a woman director, and 97 percent of these programs have a female majority on their board (Mead, 1994). This means two things: (1) programs for women and girls are controlled by women and girls; and (2) programs for women and girls are also an arena for women and girls to develop and exercise their leadership abilities.

Single-Sex Programs Can Be More Democratic

What I have discussed above offers a framework for understanding why an individual program needs to incorporate gender in order to be fully effective for all its constituents. In addition, there is an overarching reason to have an integrated set of programs that works well for women: within such an integrated system a strong democratic potential is found. A concern of many funders is that programs specifically for women are somehow antidemocratic and contradict the ideal that everyone should be welcome in every program and able to succeed in that program. In fact, the opposite may be true. Anne Schneider and Helen Ingram (1993) argue that in a truly democratic society every group would be, at one time or another, the clearly deserving recipients of public policy benefits. In looking at who benefits from public policies, Ingram and Schneider posit the existence of four groups: (1) the advantaged, those politically powerful and socially acceptable groups that typically benefit from public policies and are regarded as deserving of the benefits; (2) the contenders, those groups that fight for public policy benefits but are not regarded as automatically deserving of such benefits; (3) the dependents, those groups with little political power that are viewed as deserving of assistance but unable to help themselves; and (4) the deviants, those groups whose behavior is (in some way) judged socially unacceptable, and for whom punitive public policies are designed.

> When the policy design process operates with reasonable fairness, different interest groups within society—including social, racial, and employment groups—are subject to different constructions when they are the targets of different policies. . . . As long as social constructions are dynamic, self-correcting forces in the system prevent any group from becoming permanently disenfranchised. When these mechanisms do not work, democracy suffers. (85)

One could argue that women and girls are seen *too often* as the dependents of public policy and seen *too seldom* as the contenders or the advantaged. One very important role for gender-specific programs is to change that public policy dynamic. An additional category must be added to the four Ingram and Schneider categories: the invisible. Invisible people are those who have legitimate public concerns but who cannot get their concerns recognized or put on the public policy agenda. For example, women died of AIDS for years without being officially recognized or being able to receive the benefits available to those with the official diagnosis. Similarly, women have been the victims of domestic violence for centuries, but it is only in the past few years that women's groups have succeeded in bringing that problem to the public consciousness and insisted there be an appropriate array of public responses.

One important function of programs for women and girls, then, is to move them along this continuum of policy beneficiaries. The first step is to move women and girls out of the invisible category: to take their problems seriously, to document them, and to define them as public problems. The second step is to ensure that women and girls are not put into the deviant category. At this point in history, society views poor women who do not work as deviant—they are deemed lazy and lacking motivation. Recently, punitive policies were designed to force women back into the workplace. Programs for women and girls play an important role in countering this perception of deviance by pointing out that women with few employable skills and children to raise are not likely to succeed in the labor market—at least not without considerable support.

The third step is for programs to move women and girls out of the dependent category. They can do this in two ways: change public misperceptions of the capabilities of women and girls, and improve the skills of women and girls so they are able to take charge of their own lives. Then programs for women and girls can make them contenders. Political organizations focused on the needs of women and girls, advocacy organizations, and networks of women and girls all serve as vehicles for women and girls to contend with other groups for the benefits of public policy. Finally, programs for women and girls can move females into the advantaged category, securing their unchallenged right to benefit from certain public policies.

What Do We Look for in Effective Programs?

In evaluating programs to determine whether they are working effectively for women and girls, funders need to look at three important considerations: *equity of access*, *equity of treatment*, and *equity of outcome*. In their useful report *What's Equal? Figuring Out What Works for Girls in Coed Settings* (1993), Girls Incorporated offers a way to think about each of these equity concerns.

First, Girls Incorporated explains their main principle for working with girls in a coeducational setting. They argue that coed programs must be gender-sensitive, not gender-blind. "Leveling the playing field is more than simply opening more doors for girls and giving equal treatment to girls and boys; it is transforming the way we look at gender as it relates to girls' and boys' development" (3). "Effective strategies for working with girls in coed settings will specifically take gender socialization into account" (3).

Equity of access means that a program provides women and girls equal opportunity with men and boys to participate in programs and activities. Programs will not necessarily achieve equity of access simply by opening the door to both genders. They must ask (and answer) several hard questions. For example, What are the subtle and overt messages that invite and encourage women and girls? What messages keep them away? Is the program located in an area where females are comfortable traveling? If the program works with mothers, does it respond to the childcare issues these women face on a daily basis?

Equity of treatment means that a program offers the same level and quality of attention and resources. Some coed programs may argue that they make equal resources available to boys and girls, but that boys take better advantage of those resources. They may need to ask whether the same treatment is enough if there are unequal groups to begin with. How do we ensure, for example, that girls have the opportunity and support to become interested and skilled in nontraditional areas such as computers and working with tools, or sports? " 'Average' girls are two years behind 'average' boys in team sports skills due to differences in informal practice opportunities" (Nicholson, 1992, 16). The result of this gap is that coed sports opportunities typically prove to be ineffective.

Equity of outcome is perhaps the most important but also the most elusive measure. To assess this, programs must ask, what is the gap between females and males in achievement, knowledge, confidence, persistence, and participation? Due to gender discrimination and inequitable treatment, girls may require more time and resources or different strategies to break through barriers and become equal contributors to society.

The Relevance of Gender in Youth Programs

Thus far I have offered a set of arguments for why gender matters in program design, implementation and leadership. In order to test out these arguments in actual programs, I spent four years studying coeducational youth programs. *What became obvious in my research was that gender matters very much*. Effective coed-youth programs were the exception: unfortunately, relatively few programs (2 of 25) acknowledged and then challenged gender norms and stereotypes. Unfortunately, most programs failed to do so—with considerable consequences for the girls.

On a typical school-day afternoon, at least 75 young people stream into the Boys and Girls Center, a youth program located in an urban neighborhood in the Greater Boston area. The Center strives to be an inviting alternative to "hanging out" on the street or going home to an empty house. It offers a myriad of activities, mostly designed to let young people have fun and expend pent-up energy they cannot release at school. While there is much to praise about the Boys and Girls Center, a closer look raises some troubling questions.

Boys cluster in the middle of one room, playing pool and ping pong, vying for a turn to play a video game and dividing up into knock-hockey teams. They pay no attention to the girls, who sit in small groups around the edge of the room, talking with each other and occasionally watching the boys' activities. In the gymnasium next door, boys play basketball on one side and wrestle each other on the opposite side. The girls sit up on the stage, again talking quietly in small groups. In this program, there are about three boys to every girl, and the experiences that girls and boys are having here are very different from each other.

At the Neighborhood House, in another area of Greater Boston, the scene is quite different: in this program, there are almost an equal number of girls and boys. On one side of the gym, eight girls and two boys play volleyball together. On the other side, a lively basketball game is being played, mostly with boys, but it is clear that the stellar player is one of the girls. In another area, girls and boys are sitting at computers, completing their homework. The computer area is run by one of the older girls in the program—the acknowledged computer expert. If anyone—male or female, young person or adult—needs help with one of the computers, they know she is the one to ask. Both boys and girls in this facility look engaged and content and there are smiles all around. Why? What makes this program, similar in aims and activities, so different from the first program in terms of how girls (and boys) participate?

Over the four-year period that I studied coed-youth programs like the Boys and Girls Center and the Neighborhood House, I observed their activities, talked to the staff and to many young people in their programs. The major conclusion I drew was disturbing: many coed programs failed to meet their own stated primary goal: to serve girls as effectively as boys. On the one hand, many programs seem to be based on false assumptions about gender differences that did not reflect an accurate assessment of the young people in their programs. Conversely, program staff often lacked an understanding about how real gender differences in the lives of boys and girls—differences based on unequal opportunities and experiences—affect their participation in these programs. Program staff failed to examine program activities to ask whether girls and boys might bring different skills and interests to their participation in those activities. Too often the result was a mismatch between the program's design and the girls' interests and concerns—a mismatch that caused girls to be marginalized, their needs to be unmet, and their potential to be unrealized.

This mismatch is no small matter for girls. About 71 percent of young people in the United States participate in some type of youth program every week and an overwhelming number (in some areas, as high as 99 percent) of those programs are coed. Most girls are participating in youth programs; virtually all in coed programs. The systematic failure, therefore, by coed programs to work effectively for girls is a major concern for anyone interested in girls and their potential.

Trends in Youth Programming Have Disadvantaged Girls

Although there have been two positive trends in youth programming in the past 20 years, the net result, unfortunately, has been detrimental to girls. The first trend is that today most youth programs focus on identifying and strengthening a wide range of positive characteristics in young people rather than operating on a deficiency model that isolates and remediates a specific negative trait. While some programs focus on physical development, some on social skills, and still others teach job-related skills, there has been a shift in focus toward the healthy development of young people and away from the remediation of any specific problem. Most youth programs today have a common goal: namely, helping each individual young person to develop her/his full potential. This universal approach of developing everyone's potential might be expected to benefit all young people, girls and boys, but it does not. The second major trend in youth programming, a shift from single-sex to coed programming, has lessened the potential benefit of this universal developmental approach for girls.

A brief history of how and why this shift occurred suggests why girls lost out. Ironically, the shift toward coed programs was initiated partly by a concern about a lack of youth programming for girls. When most programming was single-sex, the great majority of programs were only for boys. To ensure equal treatment, formerly all-boy programs opened their doors to girls. But there was also a cost-efficiency concern that motivated the move toward coed programming. Those who funded youth programs and those who ran them were concerned about the need to provide the best programming with limited resources. To many of these people it made no sense to have *separate* programs for girls and for boys with equivalent facilities. Both demands that youth programs operate more equitably and cost-efficiently could have met by a collaborative strategy between single-sex programs.

But once the staff of all-boy programs realized that they could also serve girls, most of them then began eliminating the all-girl programs. Programs that chose this strategy then gained the advantage of becoming the primary youth-serving agency in their area.

The history of the Boys Clubs and their transformation to Boys and Girls Clubs of America is illustrative. The Boys Clubs could have pursued a collaborative strategy with the Girls Clubs since each organization brought to the table a strong record of effective programming for girls or boys. The two entities could have merged in a way that would have allowed their respective strengths to manifest. Instead, Boys Clubs went to court to win the right to their new name and to force the Girls Clubs to change their name to " Girls Incorporated." While Girls Incorporated continues to be a strong national organization providing services to girls, virtually all coed programming, however, is now provided by a formerly all-boys agency.

This history is important because it exemplifies the way in which much of today's coed-youth programming is pasted on top of a formerly all-boys model, and my research shows the results. Most of the coed programs I studied place the needs and interests of the boys first, are better designed for boys, and are more popular with boys. Program participation rates are also revealing. Both a study of youth programs in New York City and a national study of youth-serving agencies found that, on average, coed-youth programs serve three times more boys than girls (New York Women's Foundation, 1996; Gambone and Arbreton, 1997).

Today, what this means is that as youth programs are converting to a more promising develop-mental model, most girls are still in coed programs that *don't* implement that model effectively for them. As the skills of young people are being developed and they are learning to regard themselves as future contributors to society with exciting options to pursue, *girls are missing out.*

How Coed-Youth Programs Fail Girls

Twelve of these 25 programs fall into the Differences Are Fundamental category. These programs assume young women and young men are *inherently different* in temperament, abilities, and interests. Typically, these programs reinforce the most traditional gender stereotypes because their program activities are designed in response to quite conventional notions about the needs and strengths of boys and girls. In these programs, young men are most often the actors and the doers, engaged in conventional "boy" activities such as sports, playing video games, working with computers, or building mechanical devices. The young women in these programs are often the watchers and the observers—assuming passive roles that do not reflect their development potential. Alternatively, they may be involved in "girl" activities like arts and crafts or socializing with each other.

The 25 programs that I studied embodied four distinct sets of gender practices. As a result, I developed categories to capture these differences, defining them by two factors. First, the assumption inherent in the program design about gender differences in girls and boys. Second, the ways that the program activities interact differently with girls' and boys' socialization and life experiences. The four categories I devised were (1) Differences Are Fundamental; (2) Males Are the Model; (3) We Are All the Same; and (4) Equal Voices. These categories operate on a continuum from least effective to most effective, but only *one* category, Equal Voices, is *fully effective for girls*. Additionally, these Equal Voices' programs fostered the healthy development of *boys* as well.

What *is* rare in these programs is finding girls and boys participating equally in any task or activity. When asked about the obvious gender differences in who does what, staff members of these programs generally responded, "That's the way girls are. We try to get them to [play basketball, build a model car, write a computer program, et cetera], but they don't want to." In almost every interview I conducted with staff in these programs, they assumed that the problem (if any) lay in the girls and their lack of interest in participating in boys' activities. They never questioned any of the structures in place that funneled boys toward one set of activities and girls toward another. Nor did they question why the boys were not participating in the so-called girls' activities. In other words, the girls were compared to the boys and found lacking.

Equal Voices' programs, at the other end of the continuum, also assume that there are significant differences between young women and young men, but that these differences are not innate and incapable of change. Rather, they locate the differences between genders in socially created meanings of gender that are limiting to both girls and boys. The two Equal Voices' programs I identified in my study recognize that young women often have unequal access to opportunity relative to young men. These programs work to make *both* genders aware of these social constraints, and, most critically, encourage active questioning of them. So, Equal Voices' programs acknowledge that there may be some need for different programs for young women and young men, but they believe these programmatic considerations are driven by differences in life experience, not inherent differences in male and female constitutions.

In these programs, I saw young women and young men participating as equals in many activities, as well as involved in activities that defy gender stereotypes. In one instance, boys agreed to carry out many behind-the-scenes tasks for a major event, while girls took on the public leadership roles. In another instance, several girls and boys acted as security monitors for a youth rally. Previously, only boys would have filled those roles: but the girls said that they were just as good as boys in defusing a potentially explosive situation. It turned out they were right.

The two programs in the middle of the continuum—Males Are the Model and We Are All the Same—assume sameness between the genders. They are an improvement over Differences Are Fundamental programs in that they offer, at least on the surface, equal opportunity to young women and young men to participate in all the activities of the programs. They are not grounded in such stereotypes as young men like to "do" and young women like to "watch." However, because neither of these program categories includes an active challenge to a male standard, they operate in ways that benefit young men at the expense of young women. They "privilege" the male experience over the female. Moreover, they don't do their best by young men either, since current ideas about appropriate masculine behavior and attitudes limit young men also. Only in Equal Voices' programs can one see these limits being fully explored. Only in Equal Voices' programs are stereotypically "feminine" ways of being valued highly enough so that both young men and young women can develop their fullest potential, choosing from the entire range of human behavior and possibility.

There are important distinctions between the two programs based on the "sameness" idea. The Males Are the Model program is based on a belief that young women can do anything young men can do. A distinctive feature of the four programs I identified in this category is that their activities are ones to which the average young man brings more skills and experience than does the average young woman. The activities themselves are also often those traditionally thought of as "masculine" rather than "feminine." (Examples included a bicycle-repair program, a housing-construction program, and a computer clubhouse.) In each of these activities, the typical young women had fewer skills when she started *as well as* the additional challenge of participating in a program that defies gender norms for girls. Young women were allowed (and often even encouraged) to participate in these programs, but they had to participate as *equals*, despite the fact that their skills were not equivalent to those of the young men. The fallout from these programs is that young women drop out at a much higher rate. They view themselves—and are viewed by other participants and even staff—as being less successful in the program. *When young women fail to do well in these programs, their failure reinforces the myth that girls really cannot do everything boys can do.*

The seven programs I identified that operate with a We Are All the Same assumption begin to acknowledge unequal opportunity for girls. They work very consciously to involve girls and boys equally and treat them identically. A distinctive feature of these programs, and one that distinguishes them from Males Are the Model programs, is that their activities are not traditionally male or female. Nevertheless, male behavior is still the gold standard and it is assumed that boys and girls have the same experiences and value the same male-defined goals. (In short, young women are encouraged to emulate young men but young men are not taught to emulate young women.) On one occasion, the young people decided to develop a set of activities to address youth violence. While everybody agreed that *gang* violence was an important component of their work, the girls could not convince the boys to see that *relationship* violence was equally important. In another example, at a youth rally planned by a We Are All the Same program, young women and young men shared the several visible leadership roles at the event. They gave speeches and moderated activities in equal numbers. Yet only the girls got involved with the less glamorous, behind-the-scenes work such as buying food and ordering supplies.

Without Careful Attention to Gender, Programs Simply Aren't Effective

In 23 of the 25 programs I studied, gender practices conflicted with the stated goal of the program to develop young women and young men to their fullest capacities. In the 12 Differences Are Fundamental programs, young men and young women were channeled into very different activities within the same program based on untested assumptions the staff made about inherent gender differences in interest and ability. (Many of these programs began as all-male.) Most activities offered by these programs therefore were once designed solely with boys in mind. The result is that there is often less for girls to do in the programs and few ways for girls to break into the boys' activities.

In the 11 programs that believe in gender "sameness" not "difference" (Males Are the Model and We Are All the Same), girls often struggle to succeed, because the hidden reality is that their programs "tilt" toward boys. In Males Are the Model programs, a "sameness" orientation masks real gender differences. The young people generally participate in identical activities with identical structures and supports. But in a career preparation program, for example, the young people were learning skills that the young men (on average) already possessed to some degree *before* they joined the program. Here, the less skilled young women were disadvantaged by identical treatment. In We Are All the Same programs, an insistence on gender "sameness" also came at a cost to the girls. Despite best efforts, these programs were unsuccessful in showing how superficial equality ignored very real underlying differences.

Youth development programs have the capacity to make a more positive and constructive contribution to maximizing the potential of girls—and boys. First, they can bring their program design more in line with actual *established* differences (or lack of differences) in ability and interests between young women and young men. Second, they can work with both genders to examine and explode limiting assumptions about appropriate gender traits and roles. Third, they can adjust their program design so both young women and young men can participate effectively in all the program activities. In this way, the programs will truly prepare young people for the multiplicity of roles they will actually encounter in their adult lives.

At one time, there were many more all-male youth programs and all-female youth programs. This was due, at least in part, to traditional and essentialist notions of gender differences and appropriate behavior. These single-sex programs for young women did allow them some opportunity to participate and develop in an arena free from competition with or comparisons to young men. Today, there are many pressures on youth programs to be coeducational. The push to make programs coed may have created superficial changes that disguise the underlying reality. Several (ostensibly) coed programs I examined are essentially designed for young men.

Appropriate attention to gender issues does not happen simply because a program for young people *intends* to benefit (or even *equally* benefit) young women and young men. Rather, in important ways, a program is shaped and constrained by the gender ideas it embodies—ideas that become self-fulfilling prophecies and develop into institutionalized practices containing gender bias harmful to girls and young women. Simply having an equal number of young women and men in the program, adding women to the (male) staff, or allowing the girls to do everything boys do, does not result in a program that benefits both genders equally. Unless a program examines its gender practices, assesses the needs and interests of young people, and analyzes the barriers to access that may exist for young women, it will probably work less well for young women than young men. And in the long run, it will also limit the development of young men.

Race, Class, Gender, and Sexuality

As the above research shows, programs need to analyze gender differences in life experiences, opportunities, and social roles if they want to be as effective for women and girls as for men and boys. *What will not work is analyzing gender in isolation from other categories of social-group membership.* In the last few years, an important new area of scholarship has arisen to analyze the interconnection of race, class, gender, and sexuality. This work has clearly shown that two or more categories of meaning exist and interact simultaneously. These categories affect individuals and groups of individuals and they are also historically and contextually located. Race, class, gender, and sexuality hierarchies are never static and fixed: they constantly undergo change as part of new economic, political, and ideological processes, trends, and events.

Race, class, gender, and sexuality are social constructs whose meaning develops out of group struggles over socially valued resources (Weber, 1998). The dominant culture defines the categories within each construct (race, class, gender, sexuality) as polar opposites (white/black; men/women; heterosexual/homosexual; et cetera) to create social rankings: good and bad. It also links these concepts to biology, to imply that the rankings are fixed, permanent, and embedded in nature. In reality, these concepts are neither polar opposites nor biologically determined; rather, they are systems of power relationships; historically specific, socially constructed hierarchies of domination. Therefore, gender as a category has no meaning apart from the categories of race, class, and sexuality. All these categories impact the life experiences and opportunities of people. We need to pay attention to the multiple and interactive effects of these social-group memberships in program design and implementation, and, therefore, foundations need to do the same in their grantmaking.

While it is beyond the scope of this chapter to provide detailed evidence of the varieties of intersections of these social-group memberships, I do want to clarify that the simultaneous existence of these categories means that foundations *cannot* adopt a single-gender lens to ensure effective grantmaking. Rather, they need to ask several questions of both their grantmaking and the programs they fund. How does gender impact the experiences and opportunities of the constituents of their grantees? How does race impact experiences and opportunities as well? And what about class and sexuality? In other words, foundations need to apply several interlocking lenses in their grantmaking. Or, more concretely, they need to ask very specifically, who are the different people involved in and impacted by the issue(s) we care about? How do gender, class, race, and sexuality shape who those people are, what their needs and interests are, and what their concerns and issues are? Finally, how can all these people be actively involved in working on the issues we all care about?

Notes

This chapter was originally published in 2001 as "Gender Matters: Funding Effective Programs for Women and Girls." The original version was substantially longer and included a rich section on international grantmaking and gender.

1. These Foundation Center numbers are, at best, only a rough approximation of the actual relative funding levels for greater Boston programs. Through our survey of funders we attempted to obtain more accurate figures about support for programs for women and girls, but too few funders were able to provide us with information about the gender composition of their grantees. The Foundation Center's data on Boston area funders is part of a national sample that seeks to represent nationwide funding patterns. The data was not intended for use on a regional basis and does not claim to be representative of the variety of funders found in our region. Since it is the only figure available I use it as a starting point in this chapter for the discussion about the relationship between philanthropic giving and gender.

References and Further Reading

American Association of University Women (1992). *AAUW report: How schools shortchange girls*. Washington, DC: American Association of University Women.
Bane, M. J., and Ellwood, D. T. (1994). *Welfare realities: From rhetoric to reform*. Cambridge, MA: Harvard University Press.
Bem, S. L. (1993). *The lenses of gender: Transforming the debate on sexual inequality*. New Haven, CT: Yale University Press.

Blumberg, R. A. (1991) *Gender, family and economy: The triple overlap.* Newbury Park, CA: Sage Publications.

Bonavoglia, A. (1989). *Far from done: The status of women and girls in America.* New York: Women and Foundations/Corporate Philanthropy.

———. (1991). *Making a difference: The impact of women in philanthropy.* New York: Women and Foundations/Corporate Philanthropy.

———. (1992). *Getting it done: From commitment to action on funding for women and girls.* New York: Women and Foundations/Corporate Philanthropy.

Browers, R. (1993). "Review of the integration of gender concerns in the work of the DAC. Theme 1 of the assessment of WID policies and programs of DAC members." Conference paper prepared for the Operations Review Unit of the Directorate General for International Co-operation of the Netherlands. Institute of Social Studies International Services, The Hague, Netherlands. Mimeo.

Broverman, I., Vogel, S. R., Broverman, D. M., Clarkson, F. E., and Rosenkrantz, P. S. (1972). Sex role stereotypes: A current appraisal. *Journal of Social Issues* 28: 59–78.

Buvinic, M., Gwin, C., and Bates, L. M. (1996). *Investing in women: Progress and prospects for the World Bank.* Baltimore, MD: Johns Hopkins University Press.

Capek, M. E. S. (1998). Women and philanthropy: Old stereotypes, new challenges. A Monograph Series. Volume One, *Women as donors: Stereotypes, common sense, and challenges*; Volume Two, *Foundation support for women and girls: "Special interest" funding or effective philanthropy?* ; Volume Three, *The women's funding movement: Accomplishments and challenges.* Battle Creek, MI: W. K. Kellogg Foundation, http://www.WFNET.ORG.

Capek, M. E. S., and Hallgarth, S. A., with Abzug, R. (1995). *Who benefits, who decides? An agenda for improving philanthropy: The case for women and girls.* New York: National Council for Research on Women.

Capek, M. E. S., and Mead, M. (2004). Funding Norm doesn't fund Norma: Women, girls and philanthropy. In Neil Carson (Ed.), *The state of philanthropy in America.* Washington, DC: National Committee for Responsive Philanthropy.

Chakravartty, S. (1991). *Far from done: Women, funding and foundations in North Carolina and the Southeast.* New York: Women and Foundations/Corporate Philanthropy.

———. (1992). *Far from done: Women, funding and foundations in Wisconsin.* New York: Women and Foundations/Corporate Philanthropy.

Clark, P. (1995). Risk and resiliency in adolescence: The current status of research on gender differences. *Equity Issues* 1(1).

Connell, R. W. (1987). *Gender and power: Society, the person and sexual politics.* Stanford, CA: Stanford University Press.

Connell, R. W., Ashenden, D. J., Kessler, S., and Dowsett, G. W. (1982). *Making the difference: Schools, families, and social division.* Boston, MA: Allen & Unwin.

Danziger, S. H., and Gottschalk, P. (1995). *America unequal.* New York: Russell Sage Foundation.
Danziger, S. H., and Weinberg, D. H. (1986). *Fighting poverty: What works and what doesn't.* Cambridge, MA: Harvard University Press.
Diaz, W. A. (1996). The behavior of foundations in organizational frame: A case study. *Nonprofit and Voluntary Sector Quarterly* 25(4): 453–469.
Eberhart, C. V., and Pratt, J. (1993). *Minnesota philanthropy: Grants and beneficiaries.* St. Paul, MN: Minnesota Council of Nonprofits.
Ellwood, D. T. (1988). *Poor support: Poverty in the American family.* New York: Basic Books.
Erkut, E., Fields, J. P., Sing, R., and Marx, F. (1996). Diversity in girls' experiences: Feeling good about who you are. In B. J. Ross Leadbeater and N. Way (Eds.), *Urban girls: Resisting stereotypes, creating identities* (53–64). New York: New York University Press.
Feminist Majority Foundation. (1991). *Empowering women in philanthropy.* Arlington, VA: Feminist Majority Foundation.
Ford Foundation. (1979). *Financial support of women's programs in the 1970s: A review of private and government funding in the United States and abroad.* New York: Ford Foundation.
Foundation Center. (1992). *Grants index for 1996.* New York: Foundation Center.
Foundation Center's Statistical Information Service. (2005). www.fdncenter.org/fc_stats.
Fry, C. (1990). "Sex related differences in mathematical achievement: Learning style factors." Paper presented at the annual meeting of the American Educational Research Association, Boston, MA, April.
Galvin, K. (1990). *Far from done: The challenge of diversifying philanthropic leadership.* New York: Women and Foundations/Corporate Philanthropy.
Gambone, M. A., and Arbreton, A. J. (1997) *Safe havens: The contributions of youth organizations to healthy adolescent development.* Philadelphia, PA: Public/Private Ventures.
Garofalo, G. (1993). *Women taking power: The quest for equality.* New York: Women and Foundations/Corporate Philanthropy.
Gibbons, J. L., Hanley, B. A., and Dennis, W. D. (1997). Researching gender-role ideologies internationally and cross-culturally. *Psychology of Women Quarterly* 21(1): 151–170.
Goetz, A. M. (Ed.) (1997). *Getting institutions right for women in development.* London: Zed Books.
Greenstein, S., and Spiller, P. (1996) *Estimating the welfare effects of digital infrastructure.* Cambridge, MA: National Bureau of Economic Research.
Henderson, K. A. (1995). Inclusive physical activity programming for girls and women. *Parks and Recreation* 30(3): 70–78.
Inter-American Development Bank. (1995). *Women in the Americas: Bridging the gender gap.* Washington, DC: Johns Hopkins University Press.

Irvine, J. M. (1994). *Sexual cultures and the construction of adolescent identities*. Philadelphia, PA: Temple University Press.

Jahan, R. (1997). Mainstreaming women and development: Four agency approaches. In K. Staudt (Ed.), *Women, international development, and politics*. Philadelphia: Temple University Press.

Kabeer, N. (1994). *Reversed realities: Gender hierarchies in development thought*. London: Verso.

Katz, M. B. (1989). *The undeserving poor*. New York: Pantheon Books.

Keating, D. P. (1990) Adolescent thinking. In S. S. Feldman and G. R. Elliott (Eds.), *At the threshold: The developing adolescent*. Cambridge, MA: Harvard University Press.

Kline, B. E. (1991). Changes in emotional resilience: Gifted adolescent females. *Roeper Review* 13(3): 118–121.

Lawrence, S., Gluck, R., and Ganguly, D. (2001). *Foundation giving trends*. New York: Foundation Center.

Leman, C. (1977). Patterns of policy development: Social security in the United States and Canada. *Public Policy* 25: 26–291.

Linn, M. C., and Hyde, J. S. (1989). Gender, mathematics and science. *Educational Researcher* 18(8): 17–19, 22–27.

Lloyd, B., and Duveen, G. (1992). *Gender identities and education: The impact of starting school*. Harvester Wheatsheaf: St. Martin's Press.

Lytle, L. J., Bakken, L., and Ronig, C. (1997). Adolescent female identity development. *Sex Roles* 37(3/4): 175–185.

Martens, R. (1988). Youth sport in the USA. In F. L. Smoll, R. A. Magill, and M. J. Ash (Eds.), *Children in sport* (17–23). Champaign, IL: Human Kinetics.

McFate, K., Lawson, R., and Wilson, W. J. (Eds.) (1995). *Poverty, inequality and the future of social policy*. New York: Russell Sage Foundation.

McIlnay, D. P. (1998). *How foundations work*. San Francisco: Jossey-Bass.

Mead, M. (1994). *Worlds apart: Missed opportunities to help women and girls*. A 1993 study of corporate and foundation giving to women's and girls' programs. Medford, MA: Lincoln Filene Center, Tufts University.

Moser, Caroline O. N. (1993). *Gender planning and development*. London: Routledge.

Mulqueen, M. (1992). *On our own terms: Redefining competence and femininity*. Albany: State University of New York Press.

National Center for Education Statistics. (1990). *A profile of the American eighth grader: NELS: 88 student descriptive summary*. National Education Longitudinal Study of 1988 (NCES 90–458). Washington, DC: Office of Educational Research and Improvement, U.S. Department of Education.

———. (1991). *The state of mathematics achievement: NAEP's 1990 assessment of the nation and the trial assessment of the states*. Washington, DC: National Center for Education Statistics, Office of Educational Research and Improvement, U.S. Department of Education.

National Council for Research on Women. (1994). Do "universal" dollars reach women and girls? *Issues Quarterly* 1: 1–5.

National Science Foundation. (1990). *Assessing student learning: Science, mathematics and related technology instruction at the precollege level in formal and informal settings. Program solicitation and guidelines.* Washington, DC: National Science Foundation.

New York Women's Foundation. (1996). *The status of programming for girls aged 9–15 in New York City.* New York: New York Women's Foundation.

Nicholson, H. J. (1992). "Gender issues in youth development programs." A paper commissioned by the Carnegie Council on Adolescent Development, New York.

Pastor, J., McCormick, J., and Fine, M. (1996). "Makin" homes: An urban girl thing. In B. J. Ross Leadbeater and N. Way (Eds.), *Urban girls: Resisting stereotypes, creating identities* (15–34). New York: New York University Press.

Piven, F. F., and Cloward, R. A. (1982). *The new class war.* New York: Pantheon Books.

Pleck, J. H., Lund Sonenstein, F., and Ku, L. C. (1993). Masculinity ideology and its correlates. In S. Oskamp and M. Costanzo (Eds.), *Gender issues in contemporary society* (85–110). Newbury Park: Sage Publications.

Rao, A., Anderson, M. B., and Overholt, C. A. (Eds.) (1991). *Gender analysis in development planning.* West Hartford, CT: Kumarian Press.

Riordan, C., and Lloyd, S. (1990). Resolved: Many students, especially women, are best served by single-sex schools and colleges. *Debates on Education Issues* 2(3): 1–7.

Ross Leadbeater, B. J., and Way, N. (1996). *Urban girls: Resisting stereotypes, creating identities.* New York: New York University Press.

Rothenberg, D. (1995). Supporting girls in early adolescence. *Eric Digest.*

Sadker, M., and Sadker, D. (1994). *Failing at fairness: How America's schools cheat girls.* New York: Charles Scribner's Sons.

Schneider, A. and Ingram, H. (1993), *How the Social Construction of Target Populations Contributes to Problems in Policy Design.* Florida State University Policy Series.

Schorr, L. B. (1988). *Within our reach: Breaking the cycle of disadvantage.* New York: Anchor Books.

Schur, E. M. (1984). *Labeling women deviant: Gender, stigma, and social control.* New York: Random House.

Servatius, M. (1992). *Shortsighted: How Chicago-area grantmakers can apply a gender lens to see the connections between social problems and women's needs.* Chicago: Chicago Women in Philanthropy.

Skocpol, T. (1991). Targeting within universalism. In C. Jencks and P. Peterson (Eds.), *The urban underclass.* (411–435). Washington, DC: Brookings Institution Press.

Spencer, M. B., and Dornbusch, S. M. (1990). Challenges in studying minority youth. In S. S. Feldman and G. R. Elliott (Eds.), *At the threshold: The developing adolescent* (123–146). Cambridge, MA: Harvard University Press.

Staudt, K. (Ed.) (1997) *Women, international development, and politics.* Philadelphia, PA: Temple University Press.

Summers, L. (1992). The most influential investment. *Scientific American* 132.

Taylor, J. M., Gilligan, C., and Sullivan, A. M. (1995). *Between voice and silence: Women and girls, race and relationship.* Cambridge, MA: Harvard University Press.

Thorne, B. (1993). *Gender play: Girls and boys in school.* New Brunswick, NJ: Rutgers University Press.

Twenge, J. M. (1997). Changes in masculine and feminine traits over time: A meta-analysis. *Sex Roles* 36(5/6): 305–325. U.S. Agency for International Development (1996). Gender plan of action. Washington, DC.

Walkerdine, V. (1990). *Schoolgirl fictions.* New York: Verso.

Weber, L. (1998). A conceptual framework for understanding race, class, gender, and sexuality. *Psychology of Women Quarterly* 22: 13–32.

Wilson, W. J. (1997) *When work disappears: The work of the new urban poor.* New York: Alfred A. Knopf.

Women Working in Philanthropy. (1990). *Doubled in a decade, yet still far from done: A report on awards targeted to women and girls by grantmakers in the Delaware Valley.* Philadelphia, PA: Delaware Valley Grantmakers.

Chapter Three

Grantmaking with a Gender Lens

GrantCraft

> You can't just paint the walls pink and call it a girls program.
> —An advocate for girls in the U.S. juvenile justice system, critiquing the superficial program designs that purport to meet girls' needs

> That's interesting. We have more boys in the public-speaking program, and our girls are involved in internal leadership. Why is that?
> —A grantmaker, recounting a turning point in a grantor-grantee conversation about gender and program design

Consciously and unconsciously, grantmakers use different "lenses" to help them understand a field, program, or organization. They might view the same landscape from several perspectives—for example, leadership, public policy, and community engagement—each time seeing something different. The lenses they choose shape their decisions.

This chapter features the ideas and experiences of grantmakers who use gender as a lens to inform their grantmaking. Working in a variety of foundations and with different program and policy interests, they come to their views on gender by different routes. For a former Peace Corps volunteer and human-rights worker, sensitivity to gender is "a matter of justice." For an anthropologist-turned-grantmaker, the fact that programs and policies that disadvantage girls actually disadvantage entire communities was an unavoidable research finding: He "didn't actually look for it," but finding it was a professional "coming of age" that has shaped his work ever since. For a grantmaker involved in global HIV research, it now seems impossible to set research priorities without understanding how the role of women within their families varies from culture to culture and affects the prospects for different public health strategies.

Whatever the reason, each of these grantmakers has accepted a simple proposition: In virtually all societies, men and women have different social positions. Their different roles and upbringings give men and women different skills, opportunities, and resources—and, usually, different amounts of power.

In this chapter, grantmakers explain how gender differences shape the prospects for effective programs and supporting social change—goals central to much of philanthropy. They explain why they choose to look at their work through a gender lens, their experiences in doing so, and the results they see. They tell how using gender as an analytic tool has transformed public health, international development, juvenile justice, and other fields, enabling those fields to serve people more effectively and contributing to a more just society. They explain, as well, how gender analysis can be used in combination with other "lenses," such as race or ethnicity, to gain perspectives on grantee organizations and their own foundations.

Gender analysis is part perspective and part practice, a way of trying to understand things and a set of techniques for converting that understanding into results. This chapter is organized to help grantmakers explore three aspects of using a gender lens:

- *Understanding it.* What grantmakers say gender analysis is and isn't.
- *Using it.* Principles and tools for examining programs and organizations through a gender lens.
- *Applying it in your own organization.* Suggestions from grantmakers who have "mainstreamed" gender analysis in their own institutions.

Where the Examples Come From

This chapter was developed through a series of interviews and informal conversations with more than two dozen grantmakers, scholars, and nonprofit practitioners in a wide array of organizations. They generously shared insights on how gender analysis has clarified unexamined assumptions, made their programs more effective, and influenced thinking and practice in their fields. At the same time, many of our contributors talked frankly about tensions they've encountered in trying to apply a gender lens in their work.

Understanding It: What Gender Analysis Is and Isn't

The techniques of gender analysis emerge from a few basic principles that help define what it is and isn't, what it does and doesn't do.

What It Is and What It Does

In describing gender analysis, grantmakers referred to four starting points: Gender is social not biological. The term "gender" is used to refer to the social positions of men and women and our assumptions about who they are. Those social differences themselves differ from society to society, place to place, and time to time. If gender were about fixed, innate, or biological differences, then we wouldn't need gender analysis at all: We would always know what differences to expect and could invent "male" and "female" programs. The point of gender analysis is to identify and anticipate differences, explore their significance, and respond to them.

Gender analysis is a form of inquiry. Gender analysis examines whether and how programs, policies, and even organizational cultures can affect men and women differently because of their different social situations. Grantmakers observe that people may mistakenly think that to use gender analysis is to subscribe to a specific ideological agenda. Yet understanding that boys and girls seem to have very different experiences of the juvenile justice system is not to say that there is an orthodox way to respond to those needs. Gender analysis frames questions; it does not dictate answers.

Gender analysis promotes social justice. The different effects of a policy, program, or institution on men and women can lead to injustices small and large. The problems extend far beyond the obvious (for example, that many political systems grant power to men but not women) to encompass subtle disparities that can produce inequitable results. It's precisely because injustices often arise from unexamined aspects of daily life that we need some sort of inquiry to understand them. When understanding leads to action, the result is often a more equitable workplace, community, or society.

Gender analysis improves programs. As grantmakers have learned from their own experience, inequitable programs or policies are often ineffective ones, as well. Programs that aspire to serve men and women, or boys and girls, often end up not working well for one or the other. Gender analysis can help identify and correct these problems.

What It Isn't and What It Doesn't Do

It's also important to be clear about what gender analysis isn't. "You say 'gender,' and they hear 'not men,' " says one grantmaker of conversations with colleagues and grantees that have gone nowhere. Grantmakers emphasized four helpful not's:

Gender analysis does not explain everything. "There's no such thing as a generic woman," points out one grantmaker. Social position is not only about gender. Class, race or ethnicity—and sometimes sexual orientation, religious affiliation, caste, and clan—matter, too. Grantmakers who use gender analysis also tend to weigh other features of social position and need. For example, it's hard to think productively about AIDS prevention for women in a generic way. The situations and needs of middle-class women in an affluent, Western country are different from those of poor women in the same country—and even further removed from those of poor women in a developing country. It's thinking about gender as one element of a social situation that can inspire new responses. As one grantmaker explained, "A gender lens strengthens what I learn from looking at the people we're trying to serve in terms of class, race, or sexual orientation."

Gender analysis doesn't compromise neutrality. The real choice is whether to engage in thoughtful gender analysis or to be guided by unexamined gender assumptions that pass for neutrality. "The heart of the matter," says one grantmaker, "is whether you're conscious and critical of the gender assumptions" that shape your thinking. To leave gender out as a consideration, she argues, is not to be neutral. It's "inherently biased toward the status quo," and since the status quo is often inclined toward male experience and perspectives, it's not neutral at all. It was that "neutral" thinking, says a former grantmaker and health researcher, that allowed people to count condoms among the anti-AIDS measures that women use, an assumption that he and others found preposterous once they stopped to reflect on it. "Of course," he says in reflecting on gender in his work, "women don't use condoms. Men do." Uncovering the fact that "neutral" really meant "male" opened the door for a new and promising strategy.

Gender analysis doesn't apply only to women and girls. Because "neutral" has in fact meant "male" in most societies, gender analysis does involve understanding the implications of policies and programs for women. But as several grantmakers argued, it should also involve assessing the needs of boys and men. "We want to support programs that offer the chance of equitable outcomes for women and girls and for men and boys," said one. As a grantmaker who works in the health area explained, there's "decent access" to health care for girls in many communities but very little access for boys. Building on that insight, one national foundation, according to its president, is "looking more at gender roles for both women and men." She notes that surveys of public attitudes increasingly find that men are dissatisfied with the roles assigned them, specially when those roles prevent them from "participating in family life."

Gender analysis is not the particular province of women. Many of the stories recounted in this guide came from men who apply gender analysis in their work. Moreover, our contributors stressed that thoughtful, deliberate grantmaking comes from listening to many voices—women's and men's—in a field or community. Regarding her own role in developing strategy, one grantmaker commented, "Just because I am a woman and from a particular ethnic group does not mean that I know what needs to be done in a community."

Using It with Grantees: Gender Analysis in Grantmaking

Using a gender lens in grantmaking can raise issues of power—and not only the obvious ones having to do with relationships between men and women, but also subtle ones having to do with relationships between grantmakers and grantees. To begin, it may be helpful to distinguish between analysis of *grantee programs* and analysis of *grantee organizations*.

Looking at Programs through a Gender Lens

In theory, looking at programs through a gender lens fits easily within the work of grantmaking because it is a method for exploring program effectiveness—something grantmakers and grantees are accustomed to doing together. But grantees may not see the connection between effectiveness and gender. To the contrary, explained one grantmaker, many people seem to think, "That was a '70s thing. Women and men are equal now. We don't have to worry about that." Consequently, the challenge for grantmakers is twofold: first, to stimulate thinking about gender and effectiveness; and, then, to use that thinking to improve program effectiveness. Here are some suggestions:

Use a standard protocol for a first look. Some foundations have developed standard protocols to help grantmakers think about the gender implications of proposed programs. A protocol can help answer the question, has this program taken gender into account? Grantmakers caution, however, that the questions can sometimes provoke a counterreaction, especially if the grantee has not given much thought to gender issues. If posed too early in the discussion, one grantmaker reports, questions about gender can lead some people to "just shut down." A checklist can sometimes be a better tool for organizing your own thoughts in advance than for engaging a grantee directly.

During discussions with grantees, encourage curiosity. The biggest resource for helping grantees improve programs isn't a grantmaker's knowledge—it's a grantee's curiosity. Uncovering the gender implications of a program, and then figuring out how to respond to them, are creative acts. According to grantmakers who regularly use a gender lens, their first challenge is often to help grantees become curious, in effect helping them move

- *From indifference* . . . If grantmakers bypass the challenge of nurturing curiosity and baldly inject gender into the discussion, grantees may not see the relevance of gender to their work, or how it can contribute to program effectiveness. Because they're not intellectually engaged, they become "somewhat perfunctory," explained one grantmaker. In the worst case, grantees simply comply with what they think the funder wants: "You get, 'OK, we'll set up a girls program,' " she explains, without asking questions that can lead to different approaches. Gender analysis is reduced to gender hoop jumping.
- *. . . to inquiry.* Grantmakers report a different result when they can stimulate curiosity, usually, as one said, by being "enthusiastic" and "doing half the learning myself." When a grantmaker's curiosity becomes contagious, grantees tend to "look more closely," says another grantmaker. In the best cases, they begin making observations about their own programs, and then reflecting on them. "You hear, 'Oh that's interesting,' " she said, recounting a typical example. " 'We have more boys in the public speaking program, and our girls are involved in internal leadership. Why is that?' " When grantees frame questions about their own work, real gender analysis begins.

Use "effectiveness questions" to uncover gender assumptions. The goal of gender analysis is not to help grantees focus on gender per se but on how gender and program effectiveness are related. Instead of talking about gender, explained one grantmaker, it may make more sense to ask about the program. Typical questions include:

- *How does the program work?* Assumptions about what we know and how things work can sometimes be assumptions about what we know about men and how things work for men. A recent example has fast become a classic. Studies of emergency room workers assessing heart attack symptoms suggest they really rely on their knowledge of men's heart attack symptoms. Since what's effective for men is not always effective for women, asking how a program might work for different users is one way of checking assumptions.
- *Where does program outreach take place?* Thinking about the location of programs can often uncover assumptions about who will be served, and how. For example, a job-training program that recruits only in welfare offices is likely to attract only women, since men aren't eligible for public assistance in many jurisdictions.
- *When are programs offered?* Reflecting on timing might also uncover assumptions. Timing affects some participants, such as mothers with young children or people who work more than one job, profoundly.

During discussions, watch for jargon creep. Grantmakers who have started reading about gender analysis and discussing it with peers often find that specialized terms become a useful shorthand. But a lot of those terms are impenetrable jargon for many grantees. One grantmaker learned this the hard way. After a long meeting in which she had stressed the need for "gender balance," one participant approached her for clarification about her concern for "agenda balance." She's watched her jargon ever since.

Another grantmaker is vigilant about the way grant seekers use jargon to frame their arguments, especially when they know a funder is interested in gender issues. "If they're really good at the language, they can run circles around you," she said. "You have to press to find out what they're really saying—and what it means for constituents."

Ask what "universal" really means. Some grantees feel that by offering a universal or coed program they have sidestepped the gender dilemma and, moreover, done the "fair thing" by offering the same program to everyone. But universal programs always rest on some sort of assumptions about gender.

The challenge is to help a grantee identify assumptions that might have been overlooked and recognize their consequences. For example, researcher Molly Mead, in examining youth development programs, found that programs that attracted few girls and marginalized them were often based on assumptions that equated a "neutral" perspective with a boys' perspective.

Think about other lenses that might apply. To understand the impact of gender, it's often necessary to bring race, class, culture, and other factors into the picture. One West Coast grantmaker recalled a project on immigrants' rights in the workplace: "When we used the two lenses together—gender and immigration status—we discovered that a lot of women in our region were working in other people's houses, taking care of children or house cleaning. As a result, we helped our grantee start a cooperative to represent and support workers employed in homes."

Use brainstorming to move from diagnosis to design. The grantmaker's role in responding to gender issues depends partly on the culture and practices of the foundation. If the foundation encourages a more hands-off approach, the grantmaker might simply urge a grant seeker to give further thought to gender issues uncovered during discussions. If collaboration and give-and-take are more the norm, moving from diagnosis to design typically involves joint brainstorming.

Recalling a typical example from his work at a family foundation, one grantmaker described how he asked the dean of a theological seminary about the enrollment of women in his school. The dean noted that very few women were enrolled, despite efforts to recruit female students. In further discussion, the two began to focus on faculty composition: Perhaps the absence of women on the faculty was affecting the program's appeal to women. They began exploring options, which ultimately led to a grant for a faculty recruitment program. "They had been thinking about this problem all along," notes the grantmaker, "but didn't think to ask for a grant about it."

Encourage experimentation and research when the best design isn't apparent. A closer look at data may reveal significant patterns. In one workforce development program, for example, data disaggregated by gender showed that women's travel patterns were different because of childcare and household responsibilities. To probe this finding, the program interviewed a sample of women, then began to offer participants a ride home after work and access to emergency transportation.

Sometimes a gender problem is clear but the solution is not, and supporting experimentation is the best grantmaking strategy. In their book, *Effective Philanthropy*, Mary Ellen Capek and Molly Mead recount one example. Grantmakers and grantees at a public foundation noted that girls' attendance at some youth programs was very low compared with boys', and that girls who did attend sometimes barely participated in the program activities on offer. Additionally, girls and boys tended to segregate themselves, with boys in the computer lab and girls in the craft room. The funder offered a grant to help the program managers experiment. Among other things, they assigned all girls to the computer room and all boys to the craft room for a short period. The experience disproved the program staff's original assumptions—that boys simply don't like crafts, and girls don't like computers—and led them to encourage kids to try both programs. When coed programs were restored, attendance and participation by girls in all program activities increased.

Support wider learning and experimentation. Sometimes grantmakers encourage experimentation among a number of grantees—at a field level. A family foundation with an interest in juvenile justice, for example, learned from research and grantee reports that programs were not meeting the needs of girls. The reports suggested broad lessons for programming, but they didn't specify best practices. To encourage the development of new programs for girls, the foundation issued a request for proposals, then awarded a series of grants to organizations that converted research lessons into actual program designs. Grantmakers can also use their convening capacity to enable grantees to learn from each other. Conferences, workshops, and other contacts can help create a learning network within a field.

What about Men and Boys?

Because women and girls have been so pervasively disadvantaged by nominally "neutral" programs and policies that were actually designed for men and boys, it may seem unnecessary to study the needs of boys and men. But as many grantmakers are learning, gender analysis is increasingly uncovering issues that affect mostly boys and men. The result, as one grantmaker put it, is that "we need to do both."

In medicine and health services, for example, the idea that women are often poorly served is widely acknowledged, although the situation is not yet remedied. Drugs that work often turn out to be drugs that work for men, and "classic" symptoms often turn out to be men's symptoms. But as researchers have begun to look at questions of health services demand, they have discovered a different gender imbalance: Men are far less likely than women to seek health care, even when they need it.

According to one grantmaker, findings about low rates of health care utilization among men have prompted a new generation of outreach programs. These programs take blood-pressure screening and health education out of institutions and into neighborhood gathering spots, county fairs, and even, in a recent effort in the United Kingdom, local pubs. A few providers have sought to attract underserved men; for example, one opened a community health center aimed exclusively at low-income men. In response to these specially tailored services, says one grantmaker, "men are coming out of the woodwork." Attracting far greater public notice is the situation of boys in the American classroom. In a series of books, articles, and lectures, a handful of researchers have argued that standard classroom practices and curricula typically don't serve boys well. As William H. Pollack, director of the Center for Men and Young Men at McLean Hospital, told *Education Week* magazine, boys "come to school already socialized in a different way." Their learning, social, and emotional needs are not accommodated in most classrooms, where the curriculum "just happens to work better for girls." As a result, many boys are disengaged, discouraged, or reassigned to special education classrooms. Instead of faulting and fixing boys, Pollack argues, we should be reconsidering how classrooms work.

These changes need not happen at the expense of girls and women. This is "not a zero-sum game," concludes one grantmaker. Ultimately, she suggests, the measure of excellent programs and institutions is that they serve men and women, boys and girls equally well. But as gender analysis suggests, that can sometimes mean serving them differently.

Looking at Organizations through a Gender Lens

For some grantmakers, a commitment to gender analysis is bound up in commitments to organizational equity and diversity. In conversations with grantees, they tend to ask questions about women's access to power and opportunity within the organization: How many women are involved? In what capacity? With what authority or influence? To grantees, these are complicated questions, and the foundation that asks them may seem like a powerhouse bent on imposing its views. The grantmakers we interviewed conclude that handling these conversations requires three things: an understanding of their own organization; a willingness to learn about the grantee organization; and careful attention to the power dynamics of grantor-grantee interactions throughout. Here's what they recommend:

Understand Your Authorization

The first gender equity question is not for the grantee but for the grantmaker: Where does my own organization stand? In other words, am I authorized to use gender equity as a major consideration in assessing grantee organizations? Most grantmakers we interviewed see this as a matter of fairness to grant seekers, who need to know when they're dealing with a foundation's institutional priorities, not just a grantmaker's personal vision.

If you do have authorization, it can be helpful to frame equity as a publicly acknowledged institutional value, not a personal crusade. As one grantmaker explained, sometimes when she broaches diversity questions with grantees, "They seem to think, 'She's doing that because she's a woman, or because she's black.' By personalizing the issue, they can discount it." Yet other grantmakers note that they are careful not to leave their personal passion out. "It's important to say, 'These are also my values,' " said one. Grantees are less apt to treat the issue perfunctorily if they see that the grantmaker is talking with authenticity and care.

If you're not authorized, suggest several grantmakers, you've got to start your work at home. Try to build your own institution's commitment to gender awareness before you bring it to grantees.

Understand Your Institution's Rationale

To present gender equity as an institutional priority, grantmakers need to understand how their foundations got to it. Why are they promoting it? Grantmakers described several strands of the case for gender equity as an operating principle, and the implications for their own work:

It's a matter of justice. Most foundations that promote gender equity through their funding do it because they believe it's right. For that reason, several grantmakers noted the importance of looking at leadership opportunities and access to power in informal networks, not just formal hierarchies. A rural grantmaker, for example, looked with only mild interest at the underrepresentation of women in volunteer fire departments, until he understood that the volunteer fire department is "an informal power structure where community business gets done. And by excluding women, knowingly or unknowingly, women have been excluded from that power structure."

Diverse organizations design more effective programs. Because the experience of living as a woman in a society is different in some ways from that of living as a man in the same society, women's perspective on program design may be different from men's. This is a central proposition in diversity efforts—that more perspectives generate more insights and innovations. The Ugandan health worker who drew on her own assessment of how anti-AIDS strategies failed women is a case in point: She saw what many men did not. Involving women, however, should not substitute for thoughtful gender analysis, nor should it relieve men of that responsibility. But it does increase the chances for gender awareness and, therefore, the chances for program effectiveness. Another grantmaker, who does not accept the idea that there is "a distinctive women's leadership—more relational, less hierarchical, and so on," does believe that "women leaders tend to change the agenda, because women often draw on different networks and bring new leadership in, and their networks tend to be much closer to the problem."

Diversity improves outreach. If mobilizing supporters is essential to a nonprofit's mission, then thinking about the organization's diversity is critical. One grantmaker recalls talking to the founders of a public foundation about raising money for its grantmaking program: "An hour and a half into the meeting, I said, 'Because you're a public foundation that will be reaching out and trying to involve the community, you might consider what it means that you have only men involved.' " A few months and two planning retreats later, the founders had assembled a much more diverse board, which positioned them to tap community resources more effectively.

Diversity improves quality of life. Some grantmakers have concluded that "organizations of all sorts are healthier and work better to the extent that women are considered." For example, workplace policies that are sensitive to the needs of working mothers—with flexible hours or good leave policies—are "good for everyone." Preventing stress or burnout by helping workers balance work and personal life might start as a way to promote women's involvement or advancement, but the policies usually benefit men and single or childless women, as well. And beyond ensuring fairness in the workplace, taking gender into account can help build a more inclusive workforce, whose members encompass a greater variety of experiences, perspectives, and talents.

Think "compatibility," not "compliance." Grantmakers concerned about their grantees' diversity expressed a dilemma. On the one hand, they believe their foundations are not only entitled to their values but have a right to look for grantees who share them. On the other hand, they are reluctant to interfere in and sit in judgment on the values of grant seekers. "We have aspirations," commented one grantmaker, "but we're not trying to manipulate people. We want them to share our values." To resolve that tension, grantmakers often position their concerns about gender equity within a broader discussion of institutional values—with the goal of allowing both grantor and grantee to make an informed decision about their compatibility. The most important part of the process is to articulate the foundation's own values. "Most of our grant applicants recognize we're concerned about equity and access," explained one grantmaker whose foundation makes its values clear. "People expect us to ask these things."

Show your institution's own struggles. Some grantmakers disclose not only their values but also their own efforts to realize them internally. Explaining why he thinks most of his grantees "wouldn't say we're heavy handed and pushy," a grantmaker from a family foundation said, "I start by talking about us and what we're trying to do in our own organization. . . . It makes it much more comfortable." Another, much bigger foundation shows grantees data-tracking changes over a number of years in the composition of its own staff and board—from a mostly white male institution to a much more diverse one. Sharing those data signals that the foundation acts on its own values and appreciates the effort required to change hiring and advancement patterns.

Look *with* grantees—not at them. Grantmakers emphasized that looking closely at diversity and gender issues is a shared endeavor. They use several techniques for joint inquiry:

- *Observation.* Some take an informal approach. "We don't ask for a written census," said one grantmaker. "The exact numbers don't matter so much. But we'll sit down and go through the board list" and talk about staff composition during site visits.
- *Benchmarking.* Whether or not they use a formal census, grantmakers often refer to the composition of the local population during discussions about diversity. "I try to make it clear that I'm thinking about their diversity with reference to the general population," explained one. "If there are organizations with 25 percent or fewer [women on staff and board], I would definitely try to explore that."
- *Using a diversity table.* Several organizations ask grantees for a written breakdown of board and staff composition. These foundations believe that a numerical reckoning like this is often the best way for an organization to start thinking seriously about its diversity. "The diversity table helps me think," says one grantmaker who uses it. "It's a basic, crude instrument. Sometimes you have a great project on paper, and then you see the table and it does make you question it." On the few occasions when grantees

have objected to what they see as "quotas," one grantmaker makes clear that the goal isn't to force any action on grantees, but "to get them to think of their boards as incredibly diverse resources and perspectives."

Develop a Timing Strategy

Timing conversations about diversity can be important—and tricky. Grantmakers have to decide whether to raise diversity issues before a grant is awarded or after, and then work with the grantee to determine how long it might take to address the problems that could be uncovered.

- *In advance . . .* Raising diversity questions early in grantor-grantee discussions signals that the grantee's commitment to dealing with diversity issues is important, and perhaps a deciding factor in the grant award. If it is a critical consideration for the foundation, says one grantmaker, the "worst thing you can do is have your assistant call a grantee at the last minute for data about diversity." It's too late for discussion and reflection, much less developing a plan of action.
- *Or later . . .* Some grantmakers raise the issue after the grant is made. Their foundations encourage diversity but do not condition initial grants on demonstrating it. Some foundations flag particular problems, such as a workforce that is drastically unreflective of the wider community, as issues they will "return to as factors to evaluate future grant recommendations."
- *Change takes time.* If the grantee commits to improve its diversity, don't underestimate the "time and space" required for the job. Many grantees "need lots of dialogue internally" to clarify their goals and plans, says one grantmaker. "Anyone who understands how organizations work," added another, "knows that trying to create too much change overnight can just backfire. It can destabilize everything."

Provide Financial Support for Diversity Initiatives

Many grantee organizations are so strapped—for both time and money—that expecting organizational change without financial support is unrealistic. You have to acknowledge that "money is involved," cautions one grantmaker. Sometimes a small grant can support self-study. For bigger organizations contemplating far-reaching institutional reforms, the process can be more involved and the costs greater. To understand how it had come to marginalize so many constituents, one antipoverty nonprofit appointed a special advisory commission to learn more about the needs of different populations. A longtime funder made grants in support of the commission's work and helped the grantee hire a consultant to design and implement a plan of action.

Help to Build an Internal Mandate

Sometimes the most useful action a grantmaker can take is to get the board of a grantee organization to start thinking about diversity issues—especially if it's the board itself that fails to reflect the wider population. One grantmaker explained that it's common to see "staff to staff" agreement—between the grantmaker and the executive director of the grantee organization—but not board support for taking action. To "get the board's attention," she sometimes writes a letter to the board explaining the foundation's concerns.

Troubleshoot

Many grantmakers reported that they work with grantees through informal troubleshooting. "When someone says, 'Look, we need help with this,' " said one, "we'll try to help." Often this takes the form of networking—offering referrals for board recruiting. One grantmaker keeps a long list of potential candidates handy for grantees who complain they can't find qualified women candidates.

Gender Analysis Tools

Some foundations use standard gender analysis tools to assist their grantmakers and grantees in the field. The documents listed here are examples of two common type, the interview protocol and the diversity table:

- The ClearSighted protocol, created by Chicago Women in Philanthropy, is a set of questions—some simple, some more

probing—designed to open up a conversation about gender with grantees. The protocol (available from Women and Philanthropy at www.womenphil.org) has been adapted and customized by a number of other organizations.
- The Agency Diversity Data Form, a diversity table used by the Hyams Foundation (www.hyamsfoundation.org), helps grantmakers and grantees understand how inclusive an organization is, in terms of both gender and race/ethnicity—and therefore where it might need to make changes in order to deliver on its objectives. The form is available as a downloadable spreadsheet.

The experience of grantmakers who have used these and similar tools suggests five "principles of practice," or ideas for making the best of a gender analysis tool:

Use it to start gender analysis, not to substitute for it. Although many grantmakers use formal protocols or tools to start or organize the inquiry that is at the heart of gender analysis, they caution against letting tools substitute for that inquiry. If grantmakers use the tools mechanically, the result is often perfunctory discussion or, worse, a compliance activity in which grantees simply look to please grantmakers. "It's a basic, crude instrument," said one grantmaker about the diversity table used by his foundation, "but it helps me think." The goal is to explore important topics, not complete a checklist.

Use it most before and after grantee discussions. In advance of meetings, grantmakers use interview protocols to prepare—to refresh themselves on important issues and questions. After grantee discussions, they may use the questions to organize and analyze what they've heard, turning impressions into a more organized set of reflections. Ideally, grantee discussions will not follow a methodical review of the questions in a protocol. Once the conversation gets going, it will often cover the most important points at hand, and grantmakers can use the protocol as backup, checking occasionally to see that important angles are being explored.

Use it at grant-renewal time. Tools or protocols can generate findings or analyses that are useful in reviewing progress and commitments. In discussions about grant renewals, for example, grantmakers and grantees can look back at issues they identified earlier and see how they have played out. Did gender-appropriate program design or outreach efforts really seem to pay off? How? Did the organization make any progress toward enhancing the diversity of its workforce? If so, how? What's an appropriate next goal?

Use it to signal commitment. Sometimes grantmakers introduce a tool early in their dialogue with grantees—but not because they want it front-and-center during discussions. Instead, it signals to grantees that gender analysis is an important institutional priority for the foundation, not just a personal interest of the grantmaker. It can also ease defensiveness, if grantmakers remind grantees that they have the same discussion with all grant seekers—and are familiar with the challenges that the tool often uncovers.

Use it in your foundation. You may want to use these tools within your own foundation to organize reflection and learning. In staff development meetings, grantmakers can share their experiences with and reactions to using the tool in the field, or review a "critical incident" in which things went particularly well or badly, or talk through the implications of including particular categories (such as gender, race, sexual orientation, class, religion, and sometimes others) in the analysis. The ensuing discussion illuminates not just issues about the tool itself, but also reflections on how grantees approach gender equity and grantee-grantor interactions more generally.

Applying It in Your Own Organizations

Many grantmakers believe gender analysis works best when an entire foundation—not just an individual grantmaker—supports its use. "You can give people some principles and it might change their work," one grantmaker points out, "but it might not be sustainable at the institutional level. They can put on this lens and do a great job, but it's really at the trustee level that change happens. It has to be integrated." So how does an individual grantmaker get gender analysis on a foundation's agenda? According to the grantmakers we interviewed, that challenge involves creating pockets of experimentation, conversation, and learning. Starting points might include

Emphasize Research and Learning

Three tools of the grantmaker's craft can be used to inquire about gender issues and thereby bring attention to them within the organization:

- *Scanning the landscape.* When grantmakers are learning about a field, they can inquire—from practitioners, researchers, and

policymakers—about how gender figures in program and policy development. By querying experts in youth development, one foundation learned that gender analysis had led to new approaches for working with at-risk girls. Although the foundation didn't start out using gender analysis, it ended up embracing it. In effect, the people they consulted in their scan put gender analysis on their agenda.
- *Designing evaluation.* Grantmakers who commission or conduct evaluations have other important opportunities to promote inquiry and discussion about gender.
- *Disaggregating data.* For instance, presenting them by gender may show different participation rates or outcomes for men and women, and trying to account for those differences will inevitably lead to gender analysis. But don't assume that evaluators will look for gender issues without being asked. One grantmaker, recalling a presentation by the evaluators of an after-school program, noted that they emphasized the importance of identity to program outcomes, "but they were considering only class and race. They never even looked at gender."

Let Grantees Speak

In some cases, grantees who learn important lessons about the effect of gender on their programs can stimulate interest within a foundation. The grantmaker's job is to focus attention on those findings. As many of our contributors pointed out, some of the best examples of gender analysis started in the field, with grantees, not with grantors. The push for microbicide development, for example, started with grassroots health workers and led to the Global Campaign for Microbicides. Similarly, in juvenile justice, researchers and girls' advocates have brought gender inequities to the attention of funders who are now supporting their work.

Look for an Institutional Rationale

Grantmakers who want to promote gender analysis can take a lesson from foundations that have embraced it: frame the issues in terms of the foundation's present culture and values. For example, a grantmaker at a corporate foundation explained that, because the parent company's customer base is primarily women and "we know who buys our product," the foundation is favorably inclined toward proposals that deal with women's issues. For another grantmaker, his foundation's commitment to broad principles of fairness is what "resonated for the board—rather than saying we're doing something special for women." It's easier to propose gender analysis as an expression of existing values rather than as a new orientation.

Making a Gender Lens Visible

How does a grantmaking institution communicate its commitment to gender analysis—and to diversity and equity, more broadly—to the public and to potential grantees? Where do those commitments manifest themselves?

- *In the foundation's Web site and annual report.* The Web site is the first point of contact for many prospective grantees, and it communicates a lot about a foundation's values and priorities. The annual report serves a similar function by highlighting past accomplishments that the foundation views as especially important. What policies and commitments do grantees see reflected in the mission statement and other text? What images represent the foundation and its grantees?
- *In standard application forms and information to grantees.* Grant guidelines and application requirements can attract and encourage grantees who share a foundation's values. One grantmaker noted that applicants often call her with questions about the "nuts and bolts" of completing the diversity table, then work their way into "a deeper conversation about the values and focus of the foundation."
- *In projects and evaluations.* The most important evidence, of course, is in the actual grantmaking. Who receives grants, and for what projects? Do evaluations employ gender, race, and other analytic lenses?
- *In site visits.* A site visit is a good opportunity to observe a grantee organization and give helpful feedback. "We can observe dynamics," explained one program officer, "such as who attends the meeting, who speaks, and their level of engagement during the conversation."
- *In public and professional meetings.* A grantmaker at a regional foundation said that she and her colleagues make it a regular

practice to raise issues of race and gender in public meetings. They often present on those topics during grantmakers' gatherings.
- *In alliances.* One way to learn more about using a gender lens and signal a commitment to women's issues is to collaborate on a project with a local women's fund. See www.wfnet.org for a list of these organizations.

Key Lessons from Grantmakers

- *Consider the basic proposition.* Men and women have different social positions; their different roles and upbringings can give them different skills, opportunities, resources, and, very often, different amounts of power. If that seems reasonable to you, consider learning more about using gender analysis in your grantmaking. Gender analysis is a way to understand how programs and organizations can unintentionally affect men and women, or boys and girls, differently.
- *Gender analysis is necessary but not sufficient.* Because social position (and therefore ability to benefit from programs and organizations) is not a function of gender alone, gender analysis is never sufficient by itself. Class, race, or ethnicity, sexual orientation, religious beliefs—these and other aspects of social position need to be given their fair weight in the development of effective programs and organizations.
- *Don't forget boys and men.* The status quo has tended to disadvantage women and girls, which is why gender analysis often focuses on understanding their needs and situations. Yet as suggested by recent developments in health care (where men use services less often than women) and education (where some researchers have raised questions about how well boys function in the typical classroom), the needs of men and boys have sometimes been overlooked as well.
- *Encourage grantee curiosity.* When you ask grantees to factor gender into their proposal development, you run the risk of having them treat gender analysis as one more hoop to jump through. Try to position gender analysis as a form of creative intellectual inquiry, then think along with grantees about how it might be important in a given program.
- *Mind the power dynamics.* If your foundation wants to encourage diversity in the organizations you fund, you have to walk a fine

line. On the one hand, you need to make clear that diversity (including gender equity) is an important value for your institution. On the other hand, you want to avoid imposing your values on grantees. The best course is to be clear about your values but recognize that even grantees who share them in principle might need encouragement, help, and time to change their organizations.

- *Listen to people in the field.* The insight that leads people to reexamine a supposedly "neutral" assumption often originates with someone working on the frontline—in AIDS prevention or after-school programming or faculty recruitment—who notices a problem. By listening well to evidence from the field, you can affirm the value of unconventional thinking, encourage the search for more equitable solutions, and be an ally for proponents of diversity within the organizations that receive your support.

Note

This chapter was originally published in 2005 as "Grantmaking with a Gender Lens" by the Ford Foundation.

Chapter Four

Bringing Funders Together to Talk about Girls: A Roundtable Discussion

The Valentine Foundation

Editors' Note

Although this document is over fifteen years old, we include it within for several reasons. One is that it provides a fascinating historical record of funders and scholars convening to talk about girls' programming at a time when girls' programming was virtually nonexistent. Secondly, many of the people involved have gone on to be experts in the field, and the document shares some of their early thinking about gender-focused programs. But most of all, we include it because the questions it raises are still as relevant today as they were when it was written. Indeed, its attention to issues of race and class make it stand out from many more current documents that identify boys and girls as unified and dichotomous categories.

The Valentine Foundation

The Valentine Foundation is a charitable foundation which makes grants to qualifying tax-exempt organizations. Grants are made for organizations or programs which empower women and girls to recognize and develop their full potential or which work to change established attitudes that discourage or prevent them from recognizing that potential. Grants will be given for endeavors to effect fundamental change—to change attitudes, policies, or social patterns. The trustees are particularly interested in innovative programs that offer a new approach.

Womens Way

Womens Way is a powerful voice for women. Through collaboration, innovation, and high standards of accountability, the Womens Way coalition gathers its strength from a shared vision of a society that . . .

- Is free from violence;
- Promotes equal opportunity;
- Challenges discrimination in all its forms;
- Fosters economic self-determination;
- Affirms women's right to control their lives

By *finding* and *funding* solutions, Womens Way works toward a goal of creating a more just and humane society.

Girls Conversation

On May 3, 1990, Valentine Foundation and Womens Way of Philadelphia sponsored a special gathering of researchers, funders, and advocates for a Conversation About Girls' Development. The impetus for this event was our sense that many funders lack opportunities to hear researchers and advocates talk about the very real unmet needs of girls, especially low-income girls, and that researchers and funders alike would welcome the chance to exchange information and share concerns. The result was a day of lively passionate discussion, infused with a sense of urgency about young women who are called upon to spend their adolescence coping with survival at the expense of growth into mature adulthood.

The core question of the day—"What do funders need to know to do intelligent funding to meet girls' needs?"—was approached through four subsidiary questions: What are girls' ways of developing an "own voice" and a positive sense of self? How do racial, economic, and ethnic factors affect this process? What kinds of programs can promote this process? How can foundations respond most effectively to this information?

Fueling the discussion were presentations on understanding girls' needs and their voices:

Race, Class and Gender
Harriette Pipes McAdoo, Ph.D. of Howard University,
Jane Ransom, President/CEO of Women and Foundations/Corporate Philanthropy, and
Carol E. Tracy, Esq., Executive Director of the Philadelphia Mayor's Commission for Women;
Self-Image/Spirituality: An Interior Look
Donelda A. Cook, Ph.D., of the University of Maryland,
Jeanne Maracek, Ph.D., of Swarthmore College, and

Jill McLean Taylor, Ph.D., of Harvard Graduate School of Education;
The Body: Questions of Politics, Pleasure and Abuse
Hortensia Amaro, Ph.D., of Boston University School of Medicine and
Elaine Kaplan, Ph.D., of Temple University;
The Mind: The Power of Critical Intellect
Peggy McIntosh, Ph.D., of Wellesley College Center for Research on Women, and
Michelle Fine, Ph.D., of the University of Pennsylvania

Development of Sense of Self in Girls

Much more research is needed on the development of self-esteem in adolescent girls, especially in girls of color, girls who are impoverished, and girls of ethnic minorities. However, current research indicates that adolescence marks a dramatic change in girls' development of sense of self. During this time, girls seem to come to believe that the things they have learned about the way relationships and the world work are wrong, or unacceptable. They doubt the validity of their own insights and conclusions and begin a process of subordinating their value choices and preferences to those of men and boys. They begin to accept the notion that self-sacrifice is the appropriate model for female decision making. The internal tension caused by this shift in thinking seems to result in girls' voluntarily removing themselves from a position of equality with boys to a position on the periphery of a male-dominated world that is not like them. Research also reveals the following:

- The definition and development of self-esteem in girls is different from boys (Taylor).
- There are cultural, racial ethnic, and economic class differences in development of self-esteem (McAdoo, Kaplan).
- Girls' sense of self is manifested and develops differently in female-specific groups as opposed to coed groups (Fine).
- The role of family and extended community in the development of girls' self-esteem differs with race, ethnicity, and economic class (McAdoo, Marecek).
- Development of self-esteem in girls cannot be viewed in isolation from larger issues of women's roles in society and societal impediments to women's growth and development (McAdoo, Maracek, Taylor, Amaro).
- Pervasive covert and overt violence against women in the US is a major negative factor in development of self-esteem in girls (Maracek, Kaplan, Tracy).

The Challenge for Foundations

Two very sobering statistics informed the discussion of foundation responses. First, in fiscal 1989, funding targeted specifically for girls and women hovered around 3.4 percent. Second, a study of the 75 largest foundations in the US revealed a dearth of women (20 percent), especially women of color (5 percent), among trustees. The ramifications of this information are startling. Half of the population is both underserved programmatically and underrepresented in financial decision making by foundations. If women develop a sense of self differently than do men, and the lives and needs of women reflect different realities, then clearly 3.4 percent gender-specific funding is inadequate. If gender, culture, and ethnicity inform decision making with varying perspectives and priorities, then the imbalance of representation among trustees must be viewed as a critical factor in "gender-neutral" grantmaking.

A third area of discussion focused on the general reluctance of foundations to make long-term funding commitments. The desire for flexibility and the ability to respond to new proposals were cited as primary reasons, as well as the conviction that the primary responsibility for funding programs of social benefit lies in the public sector.

The challenge for foundations is to become advocates for girls, especially girls of color, girls of ethnic minorities, and girls who are poor. Foundations can begin that task by:

- Committing more funds for girl-specific programs;
- Increasing the number of trustees who are women, especially women of color and ethnic minorities;
- Actively exploring the possibilities of comprehensive sequential, sustained collaborative funding among foundations.

Futhermore, private foundations need to consider the potential of funding critical studies of current public policies, reconceptualizations of public sector financing strategies, and the identification and coordination of already available resources.

In the final decade of the twentieth century, private foundations are both more critical and less adequate to meeting the challenges that face our society. Due to their substantial social and economic status, they represent a major, powerful, and respected voice for legitimating the needs of low-income adolescent girls and for demanding much more respect and attention for them. The resources of private foundations alone, however, are not sufficient. The larger task lies with legislative bodies that must exercise corporate responsibility, addressing in concrete programmatic and financial terms problems that affect the life and health of the entire society.

What do funders need to know to do intelligent funding to meet girls' needs? Girls are specifically different, and need specific programming. Programs for girls are underfunded. Advocates for girls are underrepresented in the grantmaking process. Poor girls are bearing the brunt of the burden of the disintegration of urban communities. They should not be left alone. They need and deserve powerful allies.

Race, Class, and Gender Considerations

Harriette Pipes McAdoo

One must be careful to avoid the ethnocentrism that is so often found in this field, as we think of programs that are effective with young girls who are members of the various groups of color within our country. Their voices will be heard more clearly if attempts are made to include the young girls themselves within the process of planning and designing of programs. We must not let our own ageism interfere with the design of programs that may be more effective than those we ourselves design. By working and learning from focus groups of young girls, we may be able to come up with more effective programs than are ever possible with a group composed only of "old women" (anyone over twenty). We know that young girls of color will have certain common traits:

- They will have a cultural background that will be rich in diversity.
- They will differ in many ways and have experiences that are different from those of white girls, regardless of social class.
- They will experience a diminution of their talents because they are from a devalued group within our society.
- They will often experience peer pressure to not be successful.
- They are victims of imbalanced sex-ratios within the community and may fall victim to sexual exploitation, especially by older males.
- They are often members of families who are in extended family arrangements of kin and friends, who are actively involved in help exchanges of goods and services.

Cultural Variables

We will be more sensitive to the cultural issues that are at play for this group of young girls if we are able to disaggregate the (1) social economic dimension from that of (2) their ethnicity. This is true for those who are Latina and African American cultures.

Too often stereotypic attitudes blind us to only looking at one segment of young women of color, those who are in poverty. We tend to ignore those of other economic groups, whose problems may not be as life threatening, but are nevertheless serious impediments to the development of positive self-concepts. We know that young girls of color will have different needs, because of the different socioeconomic groups that exist within our communities. We must look separately at the needs of black girls who are from families who are middle class (15 percent), who are working class (50 percent), and who are in poverty (30 percent).

Middle-Class African American Girls

Strengths

- Security provided by the resources of the family and churches.
- Able to move in and out of dominant group activities and functions.

Problems

- Are often ignored in the planning of programs.
- Often under intense pressure from home to succeed.
- Often overprotected by parents.
- Often under intense pressure to be adaptable to white norms. These norms may denigrate the cultural dimensions of their lives.

Working-Class African American Girls

Strengths

- They often have a realistic view of the world of work. They are aware of racism, sexism, and classism that exist today.
- They are often under the protection of active involvement in their churches and the values that they provide.

Problems

- They exist in homes with stresses that are caused by financial pressures.
- Both parents must be employed outside of the home.
- Some families have only one consistent parent.
- They experience insufficient supervision, and are thus open to experimentation in drugs, and sexual activity.

Impoverished African American Girls

Strengths

- They are often very involved in kin help arrangements. These involvements are necessary for survival. But because of the reciprocal obligations of these help patterns they often are detrimental to those who are most able.
- Those who make it must motivate themselves, but they need external help in order to actualize their goals. These girls may find mentors and role models within the community outside of their immediate families.

Problems

- Families are economically unstable and parents are overwhelmed.
- They often have low levels of supervision and protection.
- They are often without models in the home of persons who are working on a daily basis.
- They are victims of poor schools. They often drop out, to later become pregnant (and not the other way around).
- They are sexually active early and do not know about their bodies, therefore they are not prepared for pregnancies or safe sexual practices.

Mentoring is a crucial method of providing assistance in motivating, assisting, and protecting the girls. Resources and help may be organized in many different ways to meet their specific needs within a community. Research is needed that looks at the specific groups of girls within the African American community. Their needs are specific and will probably call for different programs. Middle-class girls may not have their needs met in groups of white girls, because of the commonalities mentioned above. Working-class girls, the majority of individuals, also have needs that are often overlooked in our usual planning and research approaches.

Jane Ransom

Women and Foundations/Corporate Philanthropy (WAF/CP) was formed in 1977 at a time in which women found themselves a slim minority within the field of philanthropy and representatives of a group—more than 50 percent of the population—whose issues received little or no recognition by foundations. Thirteen years after our founding there are far more women in philanthropy—in a majority of the professional positions. This is unquestionably a tremendous advance. Yet, WAF/CP has chosen "FAR FROM DONE" as its slogan for the 80s because that is what we are. We really cannot be lulled into thinking that philanthropy has now adequately and fairly addressed the needs of women and girls just because there are more skirts in foundations' offices.

In 1979, the Ford Foundation released a study which showed that only 2.9 percent of all philanthropic dollars supported programs for women and girls. Last year WAF/CP looked at grantmaking in the 80s and found that a decade later, grants to women and girls had increased by a whopping .5 percent up to 3.4 percent of all philanthropic dollars. We examined a universe of 500 foundations responsible for 45 percent of all annual giving, analyzing all of their grants of $5,000 and over. The focus was on grants specifically to women and girls as opposed to more generic grants that benefit women and girls. Some of the trends were

- During the 80s there was a near tripling in the dollar amount allotted for female health and this was mainly in the area of reproductive health (17 percent–25 percent).
- Although there were small increases in absolute dollars contributed to social welfare and education dollars, there was actually a decrease in total percentage of dollars given in these areas.
- Funding for science programs for women and girls decreased from 5.6 percent to 3.3 percent of dollars given to this population.
- Funding in the areas of economic equity, higher education, and criminal justice increased moderately.
- Leadership development continued to be a poorly funded area.

The dramatic increase in funding for reproductive health, as opposed to other aspects of women's lives, is somewhat problematic to me. On the one hand, it's disturbing that the control of women's and girl's reproductive lives is more important than their empowerment in other spheres like education and employment. After all, wouldn't greater access in these areas give us greater power over our reproductive lives? On the other hand, right now there is a dropping off of certain funding for reproductive health—particularly for adolescent pregnancy programs—in favor of a more holistic approach which views teen pregnancy as part of a range of youth problems including substance abuse, crime, and dropping out. I worry there's a danger of losing hard reproductive rights services with this approach.

Concurrent with and related to the dearth of funding for women, and also for minority groups, is a question of power which we have begun to look at this year. Who sits on foundation boards? Who are the CEOs? Who sets the parameters, makes the final decisions, and charts the future direction of philanthropic institutions?

In April 1990 WAF/CP released a report entitled "Far from Done: The Challenge of Diversifying Philanthropic Leadership." In this study, we looked at 75 foundations which control one-third of $111.4 billion of all foundation assets. These foundations were the 25 largest in each of three categories: community, private, and corporate. We studied their governing boards by race/ethnicity and gender. We also examined information on gender and racial makeup of foundation boards which is updated annually by the Council on Foundations of a larger universe—its entire membership of over 600 foundations.

Who decides where the money goes? Here are some of the things we found. Of the 75 foundations we studied, 23 had no women trustees and 30 of the 66 which responded had no people of color as trustees. Women of color made up 5 percent of the trustees on these boards. The boards of the foundations we studied are 80 percent male and 86 percent white. The boards of the COF's larger universe are 71 percent male and 96 percent white. The more you break this picture down, the more troubling it becomes. For example, there are no Asian women on any of the corporate or community foundation boards and only one on one of the private foundation boards that we studied. There are three Native Americans—all on the board of one foundation. Of the corporate foundations we looked at, 15 of the 20 respondents had no people of color on their boards and 13 had no women.

WAF/CP believes that the inclusion of the many cultures and groups which form our society in the leadership of philanthropy will enhance grantmaking and help to develop funding strategies that are more responsive to the needs of low-income women.

Carol Tracy

As director of the Mayor's Commission for Women in Philadelphia, I have worked almost exclusively on the problems of substance abuse and women. The crack epidemic has taken an unprecedented toll on young mothers and their families. As I have talked with the women affected by the crisis directly, worked with counselors in drug and alcohol programs, and reviewed the literature about this problem, several important issues have emerged that are relevant to our discussion about the needs of girls, particularly poor girls from inner-city racial minorities.

The most startling information I have learned is the correlation between childhood sexual abuse, frequently incest, and other forms of dysfunction, particularly substance abuse, in adult women. The scant literature that exists about substance abusing women documents this, and the anecdotal information from the new women-centered substance abuse programs in Philadelphia overwhelmingly supports this. In addition to sexual abuse, these women's lives are also filled with other forms of physical violence, which has lasted from childhood through adulthood.

Research and anecdotal reports on women in prison, women who are homeless, and women who have eating disorders also show significant involvement with childhood sexual abuse. It appears to me that the only taboo about incest is talking about it—the practice appears to be alive and flourishing and is having devastating consequences for adult women. The unanswered question concerning childhood sexual abuse, rape, and domestic violence is, of course, whether the incidents are increasing, or whether instead the number of reports is rising because there are now more sympathetic avenues for reporting and getting services.

It is well established in the literature that the physical and sexual abuse of girls and women affects women and girls of all races and social classes. It is also true, however, that the crack epidemic is ravaging inner-city African American and Latina women, so something beyond abuse is taking place. Obviously poverty commands and intensifies other psycho-social problems. Crack is also cheap and highly addictive, and gives feelings of power and confidence, making it particularly appealing to women and poor people who rarely experience such feelings.

Some high school girls gave me insight into something else that might be highly relevant to the current crisis among women. I spent an afternoon recently with a small group of African American and Latina high school girls. We talked about college, job opportunities, relationships, children, marriage, et cetera, the normal fare one would expect from this age group. Two issues emerged that shocked me. The girls talked about the extraordinary amount of harassment they receive from boys and men on the street. They frequently need to have some of their male classmates walk down the street from the school with them. They were more spirited about this issue (which I introduced) than any other topic. They also indicated that they had never before discussed this among themselves. The other issue is their lack of female friendships; they indicated that they did not like or trust other girls (one girl vehemently disagreed, saying that her three girlfriends were very important to her). Several indicated that their best friends were their mothers.

Historically, female friendships and networks have operated as extremely important support systems. I can only wonder what the absence of this means in the lives of women and girls and how widespread it is. I am fascinated to know whether the lack of female friendships is a widespread phenomenon and, if so, what the causes are for the breakdown of this important social network. Perhaps it is homophobia—although their reactions and comments to my questions did not seem to indicate that this is a major problem for them. It may be identification with the oppressor in that the large amount of women-hating in this country has permeated their self-esteem and relationships. It also could be that the sexual harassment and violence surrounding their lives may make them feel so powerless that they believe that they need male protection and that other women are useless to them. And finally, it could be simply the shortage of men—so many of their male counterparts are killed or imprisoned.

Several of the new women-centered drug programs in Philadelphia where mothers live with their children spend significant efforts in teaching these young women to connect and support each other. It now occurs to me that this may be the first time in their lives that they have experienced giving and receiving support from other women. We may be underestimating to an extraordinary degree the isolation that exists among women. We obviously need to know more. We need to talk more to girls, we need to listen more, and we need to help them find their voices about issues they may regard as normal and inevitable.

Self-Image/Spirituality: How Young Women Feel about Themselves

Donelda A. Cook

The perspective of most interventions for adolescent girls is much too simplistic; for example, it is not enough to focus interventions on "just saying no" to sex and drugs. Interventions must (1) identify the reasons that young women are vulnerable to drug use and premature sexual activity and pregnancies; (2) explore their affective responses to their life circumstances and teach them appropriate methods for expressing their affective responses; (3) teach them effective decision-making and problem-solving skills; (4) address the social implications of "just saying no," that is, how to refrain from destructive activities and remain in a social peer group; and (5) include application and follow-up of the behaviors and attitudes taught in the interventions, to the home and social environment; equipping young women with skills and returning them to environments which reinforce the alternate behaviors puts them "at risk" for failure.

Identity formation is crucial to the development of young women's voices. Programs which help to develop positive African and Latino and womanist identities are important in introducing girls to the "power" of their ethnic and gender heritages, which influences their own expressions of their identities. Mentoring programs should go beyond linking girls with role models. The programs should be structured to provide: (1) educational experiences; (2) opportunities for meaningful discussions about the life situations which they are encountering; (3) opportunities for enhancing their interpersonal skills; (4) participation in aspects of the mentor's social and professional world; and (5) opportunities for the girls, their mothers, and mentors to interact so they can observe that their mothers have something to offer and are valued and they can experience the essence of true sisterhood. Mentoring programs should provide training for the mentors and ongoing evaluation of the program.

It is important that programs are based on a sound theoretical perspective. Rather than deciding from a generation away, as most principle investigators are, what is best for young women, programmers should conduct needs assessments of the actual problems and current issues for the population being served. Young women are more likely to remain in programs which have inquired and responded to their own voices regarding their needs.

Jeanne Marecek

Inadvertent race, class, gender, and ethnic bias thread through many programs for girls from low-income ethnic minority backgrounds. (Given the way that issues are framed and interpreted by the mass media, how could it be otherwise?) For instance, some efforts to bolster the self-esteem/autonomy/personal efficacy of girls ignore the social, and economic realities that girls from low-income backgrounds face. The promise that all options are open ignores institutional, social and economic constraints. The message that success is simply a matter of personal effort is hollow at best and a cruel hoax at worst. (One important example is the popular idea that early childbearing is the only obstacle to middle-class for inner-city teenagers.)

Assumptions about family and community drawn from a white, middle-class perspective can lead us to overlook the strengths and resources of other social groups. Overlooking churches as resources for social change and personal growth is one example. Viewing separation from and antagonism toward one's parents as universal aspects of adolescent development may also be an inaccurate generalization. Such a view may lead programs to overlook ways of strengthening mothers' capacities to support their daughters. (Recently some family therapists have challenged the uncritical use of the concept of mothers' overinvolvement and enmeshment. Are the prevailing norms of separation and emotional distance necessarily the best way for all families to be organized?)

A reality of the existing sex-gender system is that girls and women often are victimized sexually. Messages to teenage girls often identify women's bodies as sites of shame, objectification, danger, and coercion. Programs that communicate alternative, complementary views of women's bodies are powerful, and of women as in control of their bodies and their sexuality are an important priority.

Participation in athletics is one way that girls can develop a sense of the power of their bodies. But the serious pursuit of athletics by girls and women is not accepted in all social groups. Stereotypes of women athletes as lesbians or as "masculine" persist, along with lesser facilities and discrimination in athletic opportunities.

The sexual harassment of young women is so ubiquitous on college campuses that we must assume it is rampant in other educational and social contexts as well. The use of sexist and heterosexist epithets by students is widely tolerated. In college classrooms, male students dismiss women's contributions; if these contributions are perceived as feminist, they may be ridiculed. Some teachers rely on sexual humor and sexist remarks to keep men's attention. Colleges have been slow to take definitive action against date rape and sexual abuses by fraternities. We must seek ways to "immunize" girls against the effects that harassment and abuse have on their self-worth and their sense of entitlement.

Jill McLean Taylor

Carol Gilligan's work has drawn attention to the way women and girls appear to follow a relational path of development centered on a developing sense of self in the responsive context of human relationships. Lyn Brown, in an analysis of narratives about relationships with a privileged mainly white, sample of girls, ages 7 to 16, has documented a connection between the development of a care voice and girls' resistance to losing what they know from their own experience. This often conflicts with societal norms of femininity.

As girls become adolescents, they appear to lose the ability to resist excluding themselves that younger girls have, and instead, begin to take on psychological and cultural conventions that oppose self and relationship, and define care in terms of self-sacrifice. The confusion caused by the tension between being a "good girl," as defined by the culture, and being selfish, stands out in the transcripts as "I don't know," and "I just . . ." preface thoughts and ideas that girls speak about. While adolescent boys come of age in a world "prepared for them" or "like them" in a very real sense, adolescent girls come up against a world that is "not like them" that does not value relationships and the knowledge of the world they have.

The research that Deb Tolman and I have been involved in, through a study with a group of adolescents who are considered to be "at risk" for early pregnancy and/or school dropout, has been paying close attention to the context of the adolescents' lives. We ask questions about knowledge of self and relationships, and we are attempting to document whether or not the kind of resistance or loss of resistance that Lyn Brown found is present in the adolescents. Research on Women Teaching Girls is being done in conjunction with our project, as women teachers are often caught in the bind of supporting individual girls but bringing their own experience of the culture into their teaching. In order to develop or implement intervention that would support resistance to mainstream and cultural stereotypes, and empower these adolescents, research that takes their experience seriously is needed.

Some other work Janie Ward and I have done in Somerville and Cambridge has included adolescents and parents from six different cultural groups. This intervention research around the needs of adolescents in terms of sex education has involved community workers, as well as researchers and adolescents and parents. We need this kind of knowledge about adolescents and adolescent development in order to structure intervention and to support adolescent girls. If we have an idea of development or developmental endpoint in mind, but have the endpoint wrong, then it seems funding may not be well spent or effective.

The Body: Questions of Politics, Pleasure, and Abuse

Hortensia Amaro

I was asked to provide some thoughts on what funders should look for and/or avoid in programming for young women. Since most of my research has been with inner-city black and Hispanic adolescent girls, I will focus my remarks on programs for this population. I will also preface my suggestions by saying that the recommendations below stem from my experience in doing research and community work rather than from a specific research project.

One important gap in funding of programs for young women is lack of funding for pilot programs which seek to formulate new models to meet the needs of young women. Governmental sources of funding are often tied to specific requirements which prohibit creativity and trial of new and innovative methods. My first recommendation is that foundations consider setting aside a portion of funding to support the piloting of new ideas which test out the feasibility and effectiveness of creative approaches to targeted programs for adolescent girls. Ideas which appear promising and feasible after this initial funding stage could be targeted for continued funding.

Many prevention of intervention programs attempt to intervene at the individual level while ignoring the social context in which adolescent girls live. Such programs tend to provide counseling or individually focused intervention for problems such as drug abuse or adolescent pregnancy. Other approaches ignore the individual level intervention and address the social context. These programs may focus on changing group norms about drinking or drug use. Prevention or intervention programs should address problems from a multilevel perspective even when the major focus of the intervention or prevention is at a specific level. For example, programs targeted at preventing the onset of drug use among adolescent girls should address individual factors (for example, attitudes, coping skills), social factors (for example, peer norms regarding drug use), and community factors (for example, drug dealing and trafficking, development of leadership and resources). Problems experienced by young women in the inner city are mostly rooted in problems experienced in the larger community. While addressing these problems requires individual intervention, programs which do not address the contributing social and community factors are likely to fail.

Many inner-city girls experience not one problem but a whole set of problems which are related. Programs which respond to this reality rather than dividing problems into different programs should be supported. For example, it is my experience that pregnancy and drug use in adolescent girls in the inner city are attractive options to girls who see few possibilities for themselves in our society. A program which builds relevant skills to increase her value in the job market, which provides positive relationships with mentors and role models, which builds a positive sense of self and leadership skills, and which provides peer relationships which support a positive self-definition, can serve as a model for prevention of adolescent pregnancy and drug use. Thus program models which have a comprehensive framework rather than a myopic vision of the needs of young women should be supported.

Elaine Kaplan

The following comments are based on research on teenage mothers ages 12–15. Adolescence is a difficult stage for low-income girls. They experience a great deal of conflict over their changing bodies and new maturity. Suddenly they confront a world that is quite different from the world they knew a year or so ago. Their bodies and emotions have undergone a change. They face issues around menstruation and sexuality. Teenage boys take notice of them and may pressure them into having sexual relations before they are ready. When they turn to people whom they have had to rely on, family and teachers, answers are not forthcoming. Some teenagers report that they find it difficult to approach their parents for information. When they do, in many instances the answers do not satisfy their need for information concerning their bodies and budding sexuality. In other instances, parents and teachers do not have the kind of information needed by these teenage girls. Many programs for adolescents are limited to general educational information that do not address the special needs of girls. Programs that do address the specific problems of teenage girls seem to focus mainly on sex education or pregnancy prevention. Young girls want programs that are sensitive to their whole adolescent experience.

It would be essential to include these girls in programs directed to them. Teenage girls could work in these programs as peer advisors and consultants with program directors. Peer groups are ideal for teaching the teenagers skills and responsibility. These programs that also address the specific problems of girls from different ethnic and racial and income groups. For example, low-income black teenage girls may live in female-headed households. Studies indicate that boys feel free to make sexual demands of girls from fatherless homes. Low-income girls report that they do not fully understand the connection between sexual relationships and getting pregnant. In addition, many of these girls report sexual abuse during their early childhood. Yet teachers and counselors in various programs dealing with teenagers do not know the extent to which this occurs nor do they understand the impact on girls when they grow older. I have also found that teenage mothers who take part in parenting programs have to deal with the racism and classism of the people who are working with them. On the one hand, these program workers want to support these methods. On the other hand, counselors tell me that they are confused by teenage pregnancy.

First, it seems to me, research on low-income black and white teenage girls needs to be funded. Based on the information, programs could be created that focus on teenage girls' issues. In addition, some programs may have to be developed that will retrain people who work with low-income teenage girls.

Mind: The Power of Critical Intellect

Peggy McIntosh

Suggestions solicited by Valentine Foundation, in anticipation of our May 3 research conversation, on what to look for and what to avoid in proposals about girls:

- Fund people or programs from which you have learned
- Fund people or programs which see themselves as continuing to learn
- Choose people or programs that are
  - Well organized and consistent in behavior
  - Connected to girls, but also alert to implications for broad public or educational policy changes
  - Unafraid of the media and competent at dissemination

Anonymity and Antirelationality

It has been well documented that public middle and high schools are structured in ways that relationships among teachers, among students, or between teachers and students are almost impossible. Needless to say, such a lack of relationships creates a context that feels to young women like rejection, abandonment, and refusal. This tension is exacerbated for young women who experience institutional policies as personally rejecting—for example, being "held back"—which a great majority of young women take to evidence that "see, I am stupid" (whereas most young men I have interviewed take this same event to be evidence that "they just didn't like me, that's all") (Fine, 1991).

There is powerful evidence which suggests that transforming public schools, or creating alternative schools, which are deeply relational, in which adult-student contact is deep and continuous, proves to be enormously satisfying and educationally effective—especially for young women.

Care Taking

In the 1990s, after many years of de-federalized social work and social programs, low-income adolescents are serving as "our" social workers. Young daughters and granddaughters, in particular, are called upon once they are old enough to assume responsibility to watch children and elderly kin, to translate in housing court, stand on line at the welfare office, wait for the electricity man. . . . The tension between doing for "self" and for "kin" is hands down win for "kin." Young women, across racial and ethnic groups prefer to define themselves as good daughters/sisters/granddaughters and mothers, rather than as good students. The latter is, unfortunately, considered selfish, and any school personnel who invites them to put self above family is likely to be ignored.

The Official and Hidden Curriculum

As Peggy McIntosh has so beautifully demonstrated, both the official and hidden curriculum are committed to hiding the achievements of women; resistant to inclusion of a feminist perspective in theoretical analyses; offended by taking the needs, desires and issues of young women (including young mothers) seriously in the classroom, social support and/or in the curriculum. These four features of public high schooling—silencing, anonymity, a refusal to deal with the "private" care taking needs of adolescents, and a sexist curriculum are sufficient to dismiss the critical mind of young women struggling to balance too much, too early and with too little support. It should be noted, in conclusion, that most "dropout prevention" programs are geared toward young men (in the name of vocational training, et cetera), leaving young women on their "own."

Note

This chapter was originally published in 1991 as "Bringing Funders Together To Talk about Girls: A Roundtable Discussion" by the Valentine Foundation.

Chapter Five

Saving Black Boys

Rosa Smith

> You see things and say "Why?" But I dream things that never were, and I say, "Why not?"
>
> —George Bernard Shaw

Defining moments generally do not come at a time of a leader's choosing but in the course of leading. Two early morning radio programs stand out in my memory as different instances when I suddenly understood how to best explain and frame my professional mission regarding urban public education.

The first epiphany occurred when I was a guest on a live radio talk show that aired during morning drive time. The discussion focused on a recently published investigative report on Title I education in the Columbus, Ohio, Public Schools. As superintendent at the time, I was called to answer questions. Referencing the harsh findings of the report, the radio host suggested surely these must be the worst of times for me.

I replied, "No, Bob, these are not the worst of times. These are the best of times. The worst of times would be if no one were talking about the challenges of public urban education. These are, in fact, the best of times because everyone is talking about our challenges, problems and possible solutions. These are, in fact, the best of times!"

Existential Moments

Leaders cannot be fearful of bad news. Instead, they must embrace bad news as providing yet another opportunity to support difficult discussions. During that radio show, I came to understand in a different way that the community audits, the frequent investigative reports, the revealing curriculum management audit, and the regular and boisterous visitors to school board meetings provided positive leverage for current and desired future work.

My second existential moment involving morning radio occurred as I awoke one day to a news program on the year 2000 report on juvenile incarceration. I listened carefully as they reported that 60 percent of the incarcerated juveniles under 18 years old were African American and mostly male. Thirteen percent were Hispanic, 8 percent were Asian, and 19 percent were white youth.

I found myself doing the math in my head as I listened. Nearly 75 percent of incarcerated youth under 18 years old were African American or Hispanic. Eighty-four percent were youth of color. My head kept spinning with the whole mental picture: African American male students composed about 8.6 percent of the nation's public school enrollment, yet they comprised 60 percent of all incarcerated youth in America! I lay in bed that morning—in February 2001—paralyzed as the reality of this information consumed me.

This overrepresentation of black youth and men in jails was not new information for me. But somehow, on that morning, I heard it in a new way. While nothing in my job description as superintendent mentioned incarcerated black youth, I understood clearly that morning that this terrible statistic—60 percent—was, in fact, in large part the result of what these youth had experienced in schools, and this was all about my job and my leadership.

I was immediately reminded of something Secretary of State Colin Powell said when he was president of America's Promise during a taped appearance at the AASA National Conference on Education. Powell said, "The education of our children is a matter of life and death."

My mission and moral purposes were redefined that morning, redefined as not simply the three district goals, but as saving the lives of one of our most vulnerable groups of students: black boys.

A New Vision

If school leaders believe that they are in the business of saving lives versus simply managing a big organization involved in teaching and learning and keeping board members happy, they would be totally different leaders.

If leaders believe their job is to save Johnny's life, then under their watch

- Johnny will not routinely and disproportionately arrive at kindergarten lacking social and educational school readiness.

- Johnny will not routinely and disproportionately attend the schools with the least resources.
- Johnny will not be routinely and disproportionately taught by teachers who are least qualified to create a positive learning environment for students most vulnerable to school failure.
- Johnny will not routinely and disproportionately be placed in special education classes.
- Johnny will not routinely and disproportionately be suspended, expelled, or arrested in schools for discipline acts that principals and staff should manage.
- Johnny will not routinely and disproportionately be assigned to the lowest level of courses.
- If school leaders believe their job is to save Johnny's life, leaders will support and demand that black boys experience reading success by third grade, and not just because of NCLB targets this year, but because
  - Jails are full of Johnnys who cannot read;
  - Johnny will find it impossible to land meaningful work if he cannot read;
  - Johnny will be unable to participate fully in civil engagement if he cannot read;
  - Johnny will be less likely as a parent to engage his children in reading if he cannot read;
  - Johnny's education is a matter of life and death, literally.

It is about 8-year-old Johnny's ability to read at and above the third grade level in the third grade, and it is about what 28-year-old John will not be able to do because he could not read well as a third grader.

Lifesaving Measures

Had Bob, the radio show host, called me to talk about the 2000 juvenile incarceration report, I would have had to admit that for black boys these are the worst of times because few policymakers, educators, business, and community leaders and school boards talk specifically about this crisis. Even today I do not hear public outrage and outcry about the underlying facts: the overrepresentation of black boys in special education (black male students being three times more likely than white students to be relegated to special education programs) and the high suspension and expulsion rates of black males (being twice and triple the rates of other students). In some major cities, fewer than 30 percent of African American boys graduate from high school with their peers.

Given the scope of the challenge facing this vulnerable population, I do not believe that the dreadful trajectory of most black boys will change unless political, educational, community, business and faith leaders make the proper education of these students the litmus test for their personal leadership.

The solution does not require a degree in rocket science. It does take imaginable leadership. We must create a professional culture that encourages, causes, and creates meaningful research as well as federal, state, and local policies, networks of diverse leaders acting on this issue, a sense of urgency and an active movement for a positive future for black boys. To this end, the Schott Foundation published a state report card on the public education of black male students; a resource intended to energize federal, state and local conversations and activism about black male students.

This report card may provide the impetus and courage for current and future superintendents to lead from the position of the student group most vulnerable to school failure on the following indicators: reading and math achievement, advanced placement and gifted and talented participation rates, retention rates, discipline referrals, in-school and out-of-school suspension rates, expulsion rates, in-school arrest rates, and cohort dropout and graduation rates. Whatever group of students, by race and gender, has the greatest frequency in these categories should be the accountability litmus test for everyone in each school district.

Reconsidered Actions

So, after three years of reflection and study on this and other subjects related to school success and failure in public education, what would I now do differently if I were to reenter the superintendency? What would I recommend to future superintendents about ensuring equity of resources and achievement for America's most vulnerable students? My strategies would include these:

- Create a public broad-based community reciprocal agreement with the community, the hiring authority (board of education or mayor), chamber of commerce, teacher organization and public policymakers to establish agreed-upon expectations concerning how we will move together on behalf of students. To reach these collaborative goals, I would hire an expert futurist strategy firm. For example, the Global Business Network can conduct a community future conference to generate the expectations and measurement benchmarks for each stakeholder group and publish semiannual progress reports to the community.

- Secure financial support from the business and philanthropic communities for gathering needed data, establishing reciprocal community and district accountability systems, and implementing long-term transparent internal and external communication strategies. Partner with the chamber of commerce or area community foundation and broker with a national funder as a model pilot for imaginable leadership.
- Build partnerships across the state to influence state leaders by increasing their understanding of the disparity in students' academic results and secure policy commitments for supporting students most vulnerable to school failure.
- Building on NCLB, create and establish an accountability system that requires improvement of the factually determined most vulnerable student group as the litmus test for improvement in all categories. Ask the Education Trust in Washington, DC, to monitor developments and report results.
- Conduct a district policy audit and, based on the findings and recommendations, create school policies that support a focus on the most vulnerable and address academic and equity realities within the district. Hire Phi Delta Kappa in Bloomington, Indiana, to conduct a thorough curriculum management audit and make quarterly progress reports to the board and the public.
- Conduct a separate disciplinary and special education audit and provide annual reports to the public with a five-year goal to reduce by 50 percent the actual number and need to place black male students in special education classes. Hire the Civil Rights Project at Harvard University to conduct the audit, recommend practices and staff development activities, and report semiannually on the progress of recommendations.
- Convene quarterly public meetings to listen to the concerns and recommendations of students. Twice each year listen and talk specifically with black male students.
- Conduct a financial audit to determine and report what, where, and how resources have been historically spent and implement a student-weighted formula for future allocations. Publish annual "follow the money" reports of school and district spending to parents and public.

- Initiate and support an early childhood education business and community engagement effort to implement a phase-in plan for creating a school system that begins at age 3. Secure federal and state grants to support the first years of the project and then place in the district budget.
- Use aggressive, progressive, and persuasive marketing and communication strategies to engage the public about the report findings, the vulnerable student groups, the manner in which all vulnerable students will benefit when management and academic systems are improved for the most vulnerable students.
- Work with the teachers' organization and agreed-upon leading experts with the goal of having all students reading at or above grade level by grade 3. Develop all teachers to be excellent K-6 reading teachers or adept as reading in the content area instructors.
- Develop algebra teachers in Title I-eligible schools using the leadership of Robert Moses of The Algebra Project.
- Find, hire, support, and promote excellent black, Latino, and other male teachers of color.

Black Boys: The Sad Facts

The most compelling case behind the vulnerability of black boys in school comes from these selected findings collected by the Schott Foundation.

Expulsions and suspensions. Consisting of only 8.6 percent of public school enrollments, black boys represent 22 percent of students expelled from school and 23 percent of students suspended.

Dropouts. Between 25 percent and 30 percent of America's teenagers fail to graduate from high school with a regular high school diploma. This figure climbs to over 50 percent for black male students in many U.S. cities.

Special education. Studies have found that black students nationwide are 2.9 times as likely as whites to be designated as mentally retarded. They also have been found to be 1.9 times as likely to be designated as having an emotional problem and 1.3 times as likely to have a learning disability. Since twice as many black boys are in special education programs as black girls, it is difficult to blame heredity or home environments as the root causes for these figures. In some metropolitan districts, 30 percent of black males are in special education classes, and of the remaining 70 percent, only half or fewer receive diplomas.

Graduation. While 61 percent of black females, 80 percent of white males and 86 percent of white females receive diplomas with their high school cohorts nationally, only 50 percent of black males do so.

Juvenile incarceration. One hundred and five of every 100,000 white males under 18 are incarcerated. That figure is three times as high for black youth at 350 per 100,000. Also, more black males receive the GED in prison than graduates from college.

Unemployment. Nearly 25 percent of black youths 16 to 19 were neither employed nor in school, according to the 2000 census, nearly twice the national average for this age group and six times the national unemployment rate.

Building Trust

The superintendent has a special role in all of this by using the above analysis to create a "can-do" culture, organizational capacity, and will for achieving a continuous and significant improved academic performance by black boys. The top leaders must build hope, trust, and confidence in the district and in the leadership by making themselves available to listen to the concerns of black male students and their parents. This relationship must be nurtured with sustained attention and care. The superintendent must lead from an understanding that education is a matter of life and death and that school leaders are in the lifesaving business.

Finally, the superintendent must try to develop a governing board that embraces its role as one of lifesavers and to help the board define its role to evaluate progress and increase public support for the most vulnerable students. Failing that, the superintendent may need to find another place to do this lifesaving work.

These are unimaginable times for about 60 percent of black boys who fail to graduate from high school. Their current realities scream out for imagined, skillful, committed, and courageous leaders—often found at the school level but not yet found often enough at the district and board leadership levels.

School success for black male students and other students most vulnerable to school failure depends on leaders willing to distance themselves from business as usual by thinking differently, talking differently, and behaving differently. For these students, it *is* a matter of life and death.

A Foundation's Plan for Black Boys

One reason I was persuaded to join the Schott Foundation for Public Education team in 2001 was the board's commitment to address the most vulnerable students in public education and make key decisions based on research and evaluation and its promise to support me in doing something I always had dreamed of doing for students.

At Schott, we have begun promising work on behalf of black boys and other vulnerable students. The school-related data we have collected on black boys is shocking. At the same time, it provides enormous opportunity to generate a revolutionary response to this largely ignored American tragedy.

Through our work on behalf of black boys, Schott seeks to support, encourage, and imagine the best strategies and possible policies to help superintendents, educators at other levels, parents, civic leaders, and others who want to redirect the school success trajectory for black boys but feel overwhelmed by the scope of the problem.

To this end, Schott seeks to

- Support public policies that make school success possible and probable for students who are poor;
- Support public policy for equitable funding of public education to ensure properly resourced schools: quality teachers, materials, staff development, facilities and accountability systems;
- Support public policy for universally accessible high quality early childhood education for 3- to 5-year-olds to enable poor children to arrive at kindergarten positioned for school success;
- Support the development of representative public policy leaders committed to influencing policy for vulnerable children and families;
- Raise awareness and support strategies to redirect the K-12 public school achievement trajectory of black boys and other vulnerable students;
- Tell the truth in ways that leverages the work of boards of education, superintendents, principals and teachers and the activism of parent and community leaders; and
- Advocate for great public schools for every child every day and everywhere.

Note

This chapter was originally published in 2005 as "Saving Black Boys" by the Schott Foundation.

Chapter Six

Including Boys in Our Conversation about Gender and Justice

Michael Reichert

The creation of the Center for the Study of Boys' Lives by a consortium of historic independent schools was fortuitous for me. A consulting and clinical psychologist with many years' experience with boys, their families, and their schools, I had nonetheless not had the chance to consider boys' lives as broadly as I had come to wish. With two sons, I felt considerable personal investment in our society's man making. Creating an organization to advance our understanding of boys' lives, I now find myself in the middle of significant and far-reaching debates about learning, justice, and positive youth development. How we see boys' lives and act to ensure their healthy and moral development are key questions for our society. The questions raised throughout this chapter are questions that not only researchers, but funders too must consider when focusing on building a healthy curriculum for boys.

In the consortium, two schools are day schools for boys, one is a boys' day school in a coordinate relationship with a contiguous girls' school, one is a coed day school that converted from a girls' school in the 1970s, and two are renowned boarding schools that became coeducational in the 1970s and 1980s after long and distinguished careers as boys' schools. Each of the schools, in short, has a considerable track record dedicated to boys' education and reasonable claim, on that basis, to success and expertise in that work. Yet, their support for the new center reflected their desire to put a finer point on that expertise. The mission of the new center, founded in 2001, was "to conduct research, encourage public discussion, and advocate on behalf of boys. Using research tools that give voice to boys' lived experiences, the center will strive to promote the widest sense of possibility and greatest hope for integrity in boys' lives" (www.csbl.org).

To assert that boys need advocacy tends to generate some controversy. Current convention holds that boys have been overly advantaged in society for generations. But recent work by gender theorists has been successful in deconstructing "monolithic masculinity," with the result that almost all men can be said to have been overlooked and poorly understood by previous generations' theorizing about men's lives. And particular groups of men, those falling outside the centralized and valorized ideal usually for reasons of race, class, orientation, or biography, can be more sorely treated. There is a great deal we simply have not understood or even noticed about how boys and men in particular influence our lives.

Much of what we do know in general is alarming. Whatever their position relative to females, boys suffer a host of hardships and dangers—in their families, communities, and schools. For example, boys' educational troubles can be summarized briefly. They are diagnosed with about 70 percent of the learning disabilities and are 6 times as likely to be labeled as suffering from ADHD. They account for almost 70 percent of the Ds and Fs given in high school and are vastly overrepresented in school disciplinary cases. It should come as no surprise then, that boys account for 60 percent of the high school dropouts and only about 45 percent of the college students. In contrast, girls have caught up in math and science, hold more leadership positions in high schools, achieve the status of valedictorian disproportionately, and have almost caught up on the SATs (Conlin, 1993; Pollack, 1998; Kindlon and Thompson, 1999; Lee and Owens, 2002; Sommers, 2002; Salomone, 2003). Recent popular articles in national media like *Newsweek* (Tyre, 2006) have suggested that these data reveal a crisis in all boys' lives and attempt to explain them as the result of some combination of schools' failure to adapt to boys' biology or "essential natures," the feminization of schools and feminists' efforts to promote girls at the expense of boys. A more careful reading of the research, however, reveals particularly troubling results by class and ethnicity among boys at the bottom of achievement outcomes (Salomone, 2003). Nevertheless, the data also suggest worrisome trends, especially for schools, which are dedicated to educating all boys.

The research on boys' behavior has been even more dramatic and perhaps more alarming than those pointing to an "achievement gap." Some years ago, in one of the schools where the authors have worked, a gender audit revealed remarkable though not uncommon disparities in discipline: demerits, the primary response of school staff to students' rule infractions, were skewed toward males at a rate of nearly 20,000 to 100! The 200 or so male students averaged 100 demerits each. Even granting that such disciplinary practices as the use of demerits may be gendered to begin with (for instance that the system cues on behaviors more likely for males), many of the infractions were truly over any line of tolerable social behavior. As one example, a popular, well-respected and usually well-mannered boy one evening sent an e-mail to all of his classmates graphically inventing a sexually explicit story about one of his female classmates, who had the misfortune to be the sister of a buddy with whom he was arguing.

Such conduct by boys in schools, in fact, reflects the far greater problem of male civility. Both in their impact on their own lives as well as on the lives of others around them, male behaviors can be hurtful. Recent studies, for example, have found that boys are more likely than girls to engage in a wide range of behaviors that increase their likelihood of disease, injury, and death. Males use more alcohol, tobacco, and other drugs, drive more recklessly, drive drunk more frequently, have more sexual partners, engage more often in unprotected sex, engage in high-risk physical activities, and are both exposed to and themselves perpetrate physical violence much more often than girls. Male adolescents report 174 percent more injuries than their female counterparts and are between 2 and 5 times as likely as females to be admitted to a hospital due to injury (Lee and Owens, 2002). Fatal injuries account for 75 percent of all deaths among 15–24 year old males (Bulhan, 1985; Cordes, 1985; Kleinfeld, 1998; Pollack, 1998; Buckingham, 1999; Kindlon and Thompson, 1999; Sommers, 2000; Kimmel, 2003; Stoudt, 2006).

Some may attribute such outcomes to what the authors of one book for parents called "The Big T," referring to a presumed hormonal imperative to take risks, be aggressive and wild (Elium and Elium, 1992). But in a recent review, Courtenay (2004) argued persuasively that disproportionately poor health outcomes for males have more to do with culture than biology and were the result of "modifiable" factors (1). Many years earlier, Waldron (1976) came to a similar conclusion from her examination of life expectancies for males and females, concluding that three-fourths of the difference could be explained by male lifestyle behaviors. More recently, the U.S. Preventive Services Task Force (1996) claimed that nearly one half of all male deaths might be prevented through changes in personal practices: how men and boys eat, drink, exercise, practice sex, drive automobiles, et cetera.

Our Curriculum for Boys

Carefully characterizing what ails men as "malaise," as opposed to "oppression," is important for some. There has been strong reaction to the "me too" nature of the men's rights movement and its unwillingness to address dynamics of power and privilege relative to women (for example, Farrell, 1993). While advocacy for women and girls has produced gains on many fronts, male domination of economic opportunities, public institutions, and personal relationships remains a central fact of social life. But we suggest that boys themselves, at least at the start of their lives, are innocent of these politics and that something quite different than what boys have historically received from their caregivers and educators is needed if we are to effect their being fit into identities steeped in violence, emotional unavailability, and dominance seeking. As New (2001) put it so compellingly, within the very institutions responsible for boys' development, "human needs are not met, they are made to suffer, or their flourishing is not permitted" (731).

That we teach boys how they can be men is clear. In his survey of the socialization of boys in a wide variety of societies, Gilmore (1990) found few exceptions to the rule that boys are channeled and pressured along carefully prescribed paths. We have tended to take this gendering of our children for granted, developing notions of "sex roles" and "socialization" to explain its necessity as a requirement for civilization. Lately, in fact, certain theorists of masculinity have complained that we are neglecting to teach boys sufficiently, calling for even more rigorous attention to the training of boys. The current popularity of male mentors and role models in the literature on parenting (Elium and Elium, 1992), education (Hawley, 1992), and in communities at large (Gurian, 1996) recalls earlier periods of embattled masculinity (Kimmel, 1996).

That schools and other cultural institutions contribute in this pressuring, even playing key roles, is also well established. Next to nuclear families, schools formatively structure how masculinity is developed, rewarded, practiced, and punished for boys. However one feels about the continuing debate over biologically and socially constructed masculinity, it seems safe to postulate the existence of a curriculum in each school which aims to teach boys how to become men. Their part in man making is, in fact, so focused and influential that some researchers have regarded the school as a "masculinity factory" (Heward, 1996, 39) or a paramount example of "masculinity-making devices" (Connell, 1989, 291).

Schools' gender curricula have been termed "hidden" with respect to girls; for boys they are perhaps even more invisible. Where we have come to challenge limiting assumptions and practices for girls, our abiding concern for the delicate nature of manhood seems to have prevented us from strongly challenging how we do things with boys. It seems so important that boys wind up with carefully selected masculine traits that we tolerate a range of behavior from them (and treatment of them) in the hope that these urgent goals are served. Rationalizations like "boys will be boys" still abound in our attitude.

Much is changing on the gender landscape, however. As we discover how our preconceptions have affected our ability to see girls clearly and proceed to root out prejudice and discrimination from our educational and family practices, what is often left exposed are the preconceptions we bring to our consideration of their brothers and friends. The growing willingness to question such preconceptions and conventions upsets some who seek to maintain the old order of relations between men and women.

But as the global economic order changes, placing a premium on workplace qualities of cooperation and relationship (Winter and Robert, 1980), the history of masculinity suggests that social constructions will also change; conservative social thinkers are likely to become even more threatened. Kimmel (1996) has chronicled how fundamentally definitions of masculinity have changed over time, even in such practices as dressing baby boys in blue or pink. There is every reason to suspect that many aspects of manhood we take for granted will also change as history further unfolds.

In this period of flux and change, new opportunities for understanding boys and the particular challenges of supporting their growth open up. Of great assistance in realizing this opportunity is the analysis of gender. Rather than stuffing boys and girls into narrowly determined, rigid, and enduring sex roles, gender theorists have argued that cultures instead provide "offers" to boys, parameters, for identity, behavior, and attitude development. Boys form their masculine behaviors and attitudes in an ongoing "state of play" (Connell, 1995), actively appropriating responses to social contexts from a set of possibilities and within certain institutional constraints. This appropriation of fashion, behavior and attitude is what we mean by gender identity; it has everything to do with what is deemed right and possible by the child. Identities reflect what is offered. Fluid rather than fixed, relational rather than intrapsychic or biologically based, gender reflects the structure of relations between men and women, adults and children, each group in the overall context of social control and privilege.

Once we understand the relative nature of what we offer boys and its relevance to the kind of society we wish to create, those in charge of key socializing experiences for children can be seen to be in a position of considerable power. We can evaluate the gender practice in our families, schools, and communities to insure that they offer boys healthy and appropriate support, guidance, and understanding. None of us is bound merely by what has been.

Among the schools comprising the Center for the Study of Boys' Lives, historic commitment to boys has come to mean advocacy for their freedom and integrity. As a researcher working within these schools, I have been in a position to listen to countless boys discuss their lives. Through this attention to boys' lived experience, we are beginning to discern the outlines of a program to guide our advocacy. I offer the following thoughts not as guidelines or a recipe, but as reflections on a work in progress, while we continue to research and evaluate our efforts with respect to boys. These lessons reflect conclusions from our practice that appear essential to the task of caring for boys.

Lesson One: Help Boys Stay Close

Cultural pressures operate in families to let boys go on their own very early, resulting at times in a virtual abandonment of male children. Both parents and boys themselves receive powerful messages from everywhere in the culture that intimacy and affection between sons and parents could undermine a boy's masculinity. Role expectations for parents and sons can cause boys to push away from their parents prematurely and unnecessarily at the same time as parents surrender to subtle pressures to let their sons go. The idea of masculinity as fragile and vulnerable to interference generates a hands-off policy toward boys. Silverstein (1994) has argued in her book, aptly titled *The Courage to Raise Good Men*, that parents, particularly mothers, should worry less about spoiling masculinity and bear in mind the fundamental needs of all children, including boys, for nurturance and closeness. More recently, the critical importance of connection for child development has generated ideas for "raising relational boys" (Dooley and Fedele, 2004).

Helping boys to stay close means thinking clearly ourselves about their needs for affection, guidance, and limits as well as helping them to resist enormous pressures and anxieties which define manhood as not needing anyone. At the same time, closeness must be understood as distinct from control or overinvolvement . All children need to be able to grow independent and to develop a full sense of themselves as individuals. This is the tightrope of parenting: encouraging independence, fostering relationship, while setting appropriate limits and policies.

We all want for our boys that they be able to establish intimate relationships as they mature. For various cultural reasons, intimacy can be a psychologically challenging task for many adult males.

Osherson (1992) has detailed the ambivalence men experience with respect to intimacy as they shuttle between loneliness and rigid counterdependence. But today, we are clearer about the ingredients for a healthy life and know that having close friends and intimate relationships are essential features. Where rugged individualism may have been necessary on the frontier, the need for cooperation and mutuality in today's workplace makes a renewed understanding of relationship and intimacy for boys an important curriculum.

Lesson Two: Enjoy and Appreciate Boys

As gender analysis has developed a critique of certain behaviors and attitudes common among men, one effect has been an ambivalence about the goodness of being male. While the culture learns how to nurture and cherish the minds of its females, there has been a tendency to blame males in general for the obstacles confronting girls and sometimes even to denigrate maleness itself, identifying maleness with the masculinity of oppression. Bednall (1995) has described a situation in Australia, where he headed a boys' school, in which legislation aimed at producing equitable treatment for girls came to mean antagonism toward boys. Such policies and reactions as these problematize masculinity and leave boys searching for solid footing. They need those who care to offer them as a baseline the rightness and goodness of being male.

Boys need clear messages that it is great to be a boy, that our society welcomes and celebrates them as males. It seems difficult to advocate for boys in this way without recalling the centuries of celebration of manhood that virtually fetishized masculinity, always at the expense of women and girls. Saying that it is great to be a boy is not saying that being male is better than being a girl, but that it is great to be what one happens to be. It is simply affirming how good it is to be alive. Both genders provide real opportunities for expression, satisfaction, and challenge.

What do we mean when we exclaim that it is great to be male? What is there to being male that is not defined in opposition to (and by exclusion of) the female? Controversy has raged on this question, ranging over the past several decades from sociobiologists like Gilder (1975) and mythopoets like Bly (1990) to profeminists like Stoltenberg (1990). Some feel they can point to "deep masculine" characteristics while others regard masculinity as a construction entirely devoted to maintaining male privilege and the unjust treatment of women. Given the "hot" nature of this question is it possible to think about boys independent of the politics of our time?

There are some things we can say. Clearly boys are different in certain respects from girls. Certainly, also, they enter a world in which their slight biological differences are amplified in systems of culturally constructed gender relations. What is essential about being male remains beyond our ability to discern uninfluenced by our own and our culture's bias. While we sort out the social from the biological, the political from the essential, however, we must offer boys something resembling unequivocal delight in their beings. Hurtful or insensitive behaviors do not have to be endorsed in order to approve of boys and to love them wholeheartedly. And while we emerge from our historic confusion, we can consider that our current uncertainty offers freedom and opportunity for self-definition. We get to decide what it means to be male.

Lesson Three: Protect Boys from Violence

There is a popular notion regarding boys that aggression and violent behavior are normal. Assuring ourselves that "boys will be boys," we have permitted harshness, intimidation, and violence to dominate the lives of most boys, from their playgrounds and schools to the athletic field. Miedzian (1991) has reviewed the historic explanations for this phenomenon, ranging from vestigial aggression to hormonal determination; she concluded that it was the social construction of a "masculine mystique" which explained it best. We make boys violent, in other words, in the treatment we accord them and the violence we expect and permit from them.

Most all men have experienced or witnessed violence. By violence, I do not mean the encouragement to try hard, to give one's all, or the confident sense that we will not break if we are physical in our play or challenges. Violence is interpersonal harm, usually instrumental, based on the idea that hurting another or threatening to hurt is a legitimate means to an end. While the violence which men perpetrate against women is systematic and deplorable, the largest group of victims of violence are other males. In our families from older siblings or from parents, at school in the recess yard, hallways or in gym class, in neighborhoods at playgrounds or randomly on the streets, violence can strike a boy. Because all boys know this, hypervigilance, a constant lookout for potential attack, often dominates their attention.

Subjecting boys to normative, systematic violence, often under the noses of those charged with their care, has obvious consequences for both men and the society. Trauma theory suggests, with respect to the growing problem of male-perpetrated domestic and street violence, that "hurt people hurt people" (Bloom and Reichert, 1998). We are compelled in our search to break the cycle of violence to notice how normal it is in boys' lives. If we wish boys to expect a world that will cherish them, we must start by offering them reasonable protection from random violence. We must also do something about the violence on TV and throughout the media, the war games in the schoolyard, and misguided, exploitative sports programs. We must resist any image of men as "cannon fodder" or any fatalistic resignation to male "nature" as inherently domineering, combative, or violent.

Parents are in a particularly powerful position to respond to those who would argue that, because biology is destiny, boys are meant to jostle and fight with each other, to taunt and bully and scapegoat. We know, because of our love for our sons and the way that we value them, that no child is expendable. We want our boys to feel safe and protected from the harsher forces of the world, at least until they are strong enough to understand what is natural and what is wrong. We can begin with our own attitudes and what we permit in our families, schools, and communities.

Lesson Four: Permit Many Paths to Manhood

Traditionally, a boy's choices regarding his life have been narrowly constrained by the society's need from its men. We have indicated to boys that they may express themselves and sample life's menu—in lifestyle, identity, vocation, and relationship—only in certain, limited ways. The overriding purpose of a man's life, and hence the determinant of a boy's choices, has been to be responsible, productive, and willing to sacrifice. In service to this dictate, all aspects of boyhood have tended to be evaluated in light of the contribution they make toward the boy's efficient pursuit of his goals. Society encourages with images and rewards—our public heroes, for example—and discourages with taboos, stereotypes, and sanctions. The experience of boys as they grow to manhood suggests that the message that there is one best way to be a man is continually reinforced.

One of the strongest influences on a boy's sense of possibility comes from his peer group. This peer group, though, as Thorne (1993) found in her study of playground life in an elementary school, does not assemble its norms in a vacuum. The code of manhood which our sons encounter among their friends and schoolmates absorbs and enforces the norms of the culture. Early research on gender found that acceptance by a peer group is even more important for boys than for girls (Maccoby, 1998). Each group has its own "way of life" (Dubbs and Whitney, 1980, 27) with shared values and pressures for conformity that rivals or transcends any other force in a boy's life (Harris, 1998). Not the static reproduction theory of earlier sociology, however, but a fluid social relations of testing boundaries and competing definitions best characterizes the peer culture. Boys, as sensitive as anyone else to society's reward structure, contest each other for dominance in a hierarchical world of winners and losers.

In an ethnographic study of boys' peer relations (Reichert, 2001), student life was found to be both contained by institutional culture—the "offers" available to boys for defining their masculinities—and continually redefining it through the contest for recognition and value which took place among a wide variety of masculine expressions. The homogenization of boys' notions of what it is to be a man occurred over the course of their careers in school, subjected to daily pressures, threats, inducements, and rewards. Jewish boys, African American boys, Catholic boys, working-class boys: all had family and cultural images and lessons about being male that were held quite dear. What they met with in the life of the school was a hierarchical response to their masculinity which compelled them to further define themselves, often by yielding public space and expression to the hegemony of a masculinity which was quite foreign to them. This "silencing" of different voices and the marginalization of subordinate masculinities robbed most all boys of permission to be themselves.

Recognizing and respecting difference is the currency of multicultural movements popular on many campuses. The roots of respect for difference will be found for boys in what respect they encounter for their own differences. An enforced submission to the rule of an oppressive masculine code will tend to perpetuate itself in relations between these boys and other groups. We must allow each boy to build his own sense of self based on a sensitive ear to himself and a careful consideration of a full spectrum of possibility. To do this we must find new ways to model for, value, and reward boys who exhibit the courage to resist mere conformity.

Being a man is not one thing—it can be many things, a function ultimately of a boy's aspirations, needs, and abilities. Good men come from many directions. What they have in common, perhaps, is a full sense of themselves that allows them to be generous.

Lesson Five: Assist Boys in Their Relationship to Women

Early in a boy's life, a separation from females takes place. Boys learn that they must distinguish themselves from girls and things feminine, must exclude girls, and establish a very different image. Developmental essentialists, whether of the sociobiological stripe or from the neo-Freudian, psychobiological school, argue that the separation matches normal gender development and that boys need to be apart from girls. Culturalists respond that we teach boys to distance themselves from girls in service to a misogynist power structure.

In any event, the result is that two very distinct cultures emerge from childhood, the outcome of very different experiences, influences, and challenges. Popular books speak of males and females as coming from different planets, the cultural separation being so complete. Ethnolinguists like Tannen (1990) are able to describe features of speech and styles of communication so different as virtually to constitute different languages.

When boys are reunited with females, the pressures are tremendous and the resources slim. Dating and adolescent interactions are steeped in the limitations resultant from the near complete separation just described. In addition to a virtual absence of opportunities to learn about girls, boys have the added handicap of having absorbed numerous derogatory images of girls. Kimmel (1990), for example, has detailed how early in a boy's life exposure comes to pornography. If anything, the culture subjects boys to these distorted and confusing images earlier and earlier. Trying to establish relationships with girls in which real contact can be made in the face of these circumstances can be an overwhelming challenge for boys.

Even more fundamentally, however, a barrier to good relationship occurs in the ignorance boys exhibit about the different social circumstances offered to girls and in their blindness to their relative privilege. Against the backdrop of pressured circumstances in their own lives, boys generally fail to notice girls' experiences of discrimination and disadvantage. Moreover, they encounter frequent messages about the "naturalness" of their relative position of dominance; in the United States, particularly, biological difference as an explanation for different treatment of men and women has recently enjoyed a boom time. Boys have a hard time grasping the notion that they are a privileged group, that they are "overvalued" (McIntosh, 1988). As Kaufman (1993) has put it, they have a "contradictory experience of power," with subjective experiences of powerlessness, deriving from their own subjugation to hegemonic masculinity, determining their relationship to others.

The rigidity and completeness of the separation of boys from girls and their blindness to the situation for girls is a variable which those who care about boys can address. Thorne (1993) found hopeful signs that boys will cross gendered boundaries where there are "openings." We can find ways to support and to extend such openings for boys. We can model relationships between adult males and females that reflect friendship and mutual respect. We can encourage contact with girls in all avenues of life and, where necessary, engineer opportunities for our boys to engage in activities with girls as their peers. And we can find ways to permit girls and boys to talk to each other about their lives and experiences, pressures, challenges, and hopes.

Boys will find this uncomfortable, particularly where the terms of the exchange are not necessarily familiar to them. In workshops organized for upper school boys and girls from neighboring single-sex schools, for example, there has been ongoing dispute over the degree of emotional expression encouraged. Boys generally cannot understand girls' freedom to express hurt and struggle; they prefer intellectual contest and debate. In just this one instance, though, as in many examples through the course of the workshop, boys find their ability to withstand some discomfort ultimately yielding a liberationist perspective. Girls, meanwhile, find it remarkable that boys in these workshops reveal selves so different from their more familiar postures.

Lesson Six: Provide Models and Mentors for Boys

There is another popular notion that boys must be instructed and initiated in the passage to manhood. Contemporary men's theorists, in fact, like Bly (1990) and Osherson (1986) have popularized a concept—"father hunger"—to describe boys' presumed need for father figures. The psychological model on which this idea rests may not appeal to all thoughtful parents today, however, especially when its result is that mothers are warned to get out of their sons' way.

But for a boy the value of a man to interact with, to question and regard and react to, seems unarguable. Masculinity for boys is not an abstract thing: it is the lessons drawn from the human society in which he operates, the conclusions from all of the day-to-day transactions the boy makes with his world. The value of a relationship with an adult male is the chance to learn firsthand the human dimension to manhood. How does a man get up in the morning, eat his cereal, create his life, correct his mistakes, love his partners? With flesh and blood interaction, notions of manhood can be placed in perspective; without it, they tend to be stereotypic and often exaggerated, absolute, and tyrannical to the boy who has no model for forging male passage.

Old notions of male identity have evolved through postmodern theory to notions of masculinity as "constitutive" (Connell, 1995), an ongoing process of choice and self-definition in the face of social constraints and openings. For boys, it helps to see a man negotiating these parameters, struggling himself and explaining the options he perceives, the choices he makes.

This lesson tends to be reduced to a social bromide, overlooking the contributions of mothers, overestimating the mere presence of a father, and altogether reproducing old patterns of patriarchal family life. And we often fail to find men who can tolerate the relational nature of these demands. But the message from boys is that they like to be around men, particularly men who can notice and enjoy them.

These lessons are ones that I have discovered from research and practice with boys; they are neither complete nor finished. They are offered as a way to initiate conversation about boys and about our responsibility toward them. One key lesson from gender studies is about our power to affect the lives of our children. Boys are in need of our thoughtful care.

References and Further Reading

Addelston, J. (1995). "Exploring masculinities: Gender enactments in preparatory high schools." Unpublished dissertation CUNY (New York).
Bednall, J. (1995). *Teaching boys to become "gender bi-lingual."* Hunting Valley, OH: University School Press.
Bloom, S., and Reichert, M. (1998). *Bearing witness: Trauma and collective responsibility.* Binghamton, NY: Haworth.
Bly, R. (1990). *Iron John.* New York: Vintage Books.
Buckingham, J. (1999). *The puzzle of boys' educational decline: A review of the evidence.* Canberra, AU: Center for Independent Studies, www.cis.org.au (accessed May 3, 1999).
Bulhan, H. (1985). Black American's and psychopathology: An overview of research and therapy. *Psychotherapy* 22: 370–378.

Conlin, M. (1993). The new gender gap. *Business Week*, May 26, 74.
Connell, R. W. (1989). Cool guys, swots and wimps: The interplay of masculinity and education. *Oxford Review of Education* 15(3): 291–303.
———. (1995). *Masculinities*. Berkeley: University of California Press.
Cordes, C. (1985). Black males at risk in America. *APA Monitor*, 9–10, 27–28.
Courtenay, W. H. (2003). Key determinants of the health and well being of men and boys. *International Journal of Men's Health* 2 (3) (January): 1–20.
Dooley, C., and Fedele, N. M. (2004). Mother and sons: Raising relational boys. In J. V. Jordan, M. Walker, and L. M. Hartling (Eds.), *The complexity of connection* (220–249). New York: Guilford Press.
Dubbs, P. J., and Whitney, D. D. (1980). *Cultural contexts: Making anthropology personal*. Boston, MA: Allyn & Bacon.
Elium, D., and Elium, J. (1992). *Raising a son*. Hillsboro, OR: Beyond Words Publishing.
Faludi, S. (1994). The naked Citadel. *New Yorker*, September 5, 62.
Farrell, W. (1993). *The myth of male power*. New York: Simon & Shuster.
Gilder, G. (1975). *Sexual suicide*. New York: Bantam.
Gilmore, D. (1990). *Manhood in the making*. New Haven, CT: Yale University Press.
Gite, L. (1985). Black men and stress. *Essence* 130 (November): 25–26.
Gurian, M. (1996). *The wonder of boys*. New York: Jeremy P. Tarcher/Putnam.
Harris, J. R. (1998). *The nurture assumption*. New York: Touchstone Books.
Hawley, R. (1991). About boys' schools: A progressive case for an ancient form. *Teachers College Record* 92(3): 433–444.
Heckler, M. (1985). *Report of the secretary's task force on black and minority health*. Bethesda, MD: U.S. Department of Health and Human Services.
Heward, C. (1996). *Making a man of him: Parents and their sons' careers at an English public school 1929–1950*. London: Routledge.
Kaufman, M. (1993). *Cracking the armor*. Toronto: Penguin Books.
Kimbrell, A. (1995). *The masculine mystique*. New York: Ballantine Books.
Kimmel, M. (Ed.) (1990). *Men confront pornography*. New York: Meridian.
———. (1996). *Manhood in America*. New York: Free Press.
———. (2003). I'm not insane; I am angry: Adolescent masculinity, homophobia and violence. In M. Sadowski (Ed.), *Adolescence at school: Perspectives on youth, identity and education*. Cambridge, MA: Harvard Education Press.
Kindlon, D., and Thompson, M. (1999). *Raising Cain*. New York: Ballantine Books.
Kleinfeld, J. (1998). "The myth that schools shortchange girls: Social science in the service of deception." Paper prepared for The Women's Freedom Network, http://www.uaf.edu/northern/schools/myth.html (accessed May 3, 2000).
Lee, C., and Owens, R. G. (2002). *The psychology of men's health*. Buckingham, UK: Open University Press.
Maccoby, E. (1998). *The two sexes*. Cambridge, MA: Belknap Press of Harvard University Press.

McIntosh, P. (1988) *White privilege and male privilege: A personal account of coming to see correspondences through work in women's studies.* Wellesley, MA: Center for Research on Women.

Miedzian, M. (1991). *Boys will be boys.* New York: Doubleday.

New, C. (2001). Oppressed and oppressors? The systematic mistreatment of men. *Sociology* 35(3): 729–748.

Osherson, S. (1986). *Finding our fathers.* New York: Fawcett Columbine.

———. (1992). *Wrestling with love.* New York: Fawcett Columbine.

Pollack, W. (1998). *Real Boys.* New York: Random House.

Reichert, M. (2001). Rethinking masculinities: New ideas for schooling boys. In W. Martino and B. Meyenn (Eds.), *What about the boys?* (38–52). Philadelphia, PA: Open University Press.

Sabo, D., and Gordon, D. F. (Eds.) (1995). *Men's health and illness.* Thousand Oaks, CA: Sage Publications.

Sadker, M., and Sadker, D. (1994). *Failing at fairness.* New York: Charles Scribner's Sons.

Salomone, R. C. (2003). *Same, different, equal.* New Haven, CT: Yale University Press.

Silverstein, O. (1994). *The courage to raise good men.* New York: Viking.

Sommers, C. H. (2002). *The war against boys: How misguided feminism is harming our young men.* New York: Simon & Schuster.

Stillion, J. M. (1995). Premature death among males. In D. Sabo and D. F. Gordon (Eds.), *Men's health and illness.* Thousand Oaks, CA: Sage Publications.

Stoltenberg, J. (1990). *Refusing to be a man.* London: Fontana.

Stoudt, B. G. (2006). "You're either in or you're out": School violence, peer discipline and the (re)production of hegemonic masculinity. *Men and Masculinity* 8(3): 273–287.

Tannen, D. (1990). *You just don't understand.* New York: Ballantine Books.

Thomas, D. (1993). *Not guilty: The case in defense of men.* New York: William Morrow and Company.

Thorne, B. (1993). *Gender play.* New Brunswick, NJ: Rutgers University Press.

Tyre, P. (2006). The trouble with boys. *Newsweek*, January 30, 44–52.

Waldron, I. (1976). Why do women live longer than men? *Journal of Human Stress* 2: 1–13.

Winter, M. F., and Robert, E. F. (1980). Male dominance, late capitalism and the growth of instrumental reason. *Berkeley Journal of Sociology* 22: 249–280.

U.S. Preventive Services Task Force. (1996). *Guide to clinical preventive services*, 2nd edition. Baltimore, MD: Williams & Wilkins.

Chapter Seven

Supporting Boys' Resilience: A Dialogue with Researchers, Practitioners, and the Media

Allyson Pimentel

Why Boys?

To improve the lives of women and girls in our society, men's and boys' lives must change as well. For over three decades, the mission of the Ms. Foundation for Women has been to support the efforts of women and girls to govern their own lives and influence the world around them. This work has been done with the awareness that the lives and futures of women and girls are interwoven with those of men and boys, and that the gender order in our society has harmful effects on all members of the human community.

A focus on boys is crucial. Boys—and the men that they become—are active participants in and gatekeepers of a rigid gender order that structures our lives, informs our public policy, and creates and defeats possibilities for boys and men, and for girls and women. Masculinity, as it is culturally constructed, puts forth a constricted, often destructive, version of boyhood and manhood that limits the full range of emotional and behavioral potential that boys inherently possess.

In March 2004, the Ms. Foundation for Women held a symposium to address and explore these issues of gender and masculinity, "Supporting Boys' Resilience: A Dialogue with Researchers, Practitioners, and the Media," where ways to support boys' resilience by helping them remain healthy, strong, and confident in the face of obstacles were explored. Leading members of the academic, media, and direct-service communities gathered to present and participate in a dialogue with an audience consisting of funders, academics, and direct-service practitioners. In a series of presentations, panel discussions, films, and breakout sessions, the presenters and attendees engaged in a challenging, complex, and sometimes difficult conversation about boys' resilience, their resistance and capitulation to culturally constructed images of masculinity, and the possibilities of giving new meaning to manhood.

This chapter is a description of that conversation and a presentation of those possibilities. It begins with a discussion of feminist reflections on boys and men as allies. It goes on to address the obstacles—both perceived and real—to the healthy development of boys, and emphasizes the need to build resistance and resilience in the face of these obstacles. Next, it outlines new possibilities for boyhood and manhood and provides a rationale and prescriptions for rethinking masculinity as constructed by society-at-large, the media, and the social science literature. It then interrogates the connection between masculinity and violence, and highlights specific strategies for breaking this link and healing the wounds it has wrought. This chapter ends with charting the remaining challenges we face in supporting boys'—and, interrelatedly, girls'—resilience.

Feminist Reflections on Boys and Men as Allies of Girls and Women

Feminist women have long promoted the development of healthy boys. As Marie C. Wilson, Ms. Foundation for Women President Emerita, noted, feminist women have understood that raising healthy boys is necessary for raising healthy girls and creating a healthy society. Susan Wefald, director of Institutional Planning at the Ms. Foundation, reminded us that adherence to narrowly defined gender roles for boys and men, as well as for girls and women, is a major obstacle to achieving women's equality. Professor Carol Gilligan of New York University and author of *In a Different Voice: Psychological Theory and Women's Development* (1982) stated that the joining of men and women is absolutely critical in our efforts to challenge the patriarchal order that divides us and to create a just society. Patriarchy, as explained by Gilligan (1982), is

> an anthropological term, describing families and cultures that are headed by fathers.[Patriarchy] is a hierarchy or priesthood in which a father or some fathers control access to truth or power or God or knowledge . . . As such, patriarchy is an order of domination, privileging some men over others and subordinating women. But in dividing men from men and men from women, in splitting fathers from mothers

and daughters and sons, patriarchy also creates a rift in the psyche, dividing everyone from parts of themselves. (7)

Gilligan described how individuals, relationships, and societies are forced toward disconnections dictated by patriarchal culture. For boys, this disconnection comes early in life when they are pressured to distance and differentiate themselves from their mothers to prove their masculinity. To the extent that masculinity is defined in opposition to femininity, boys learn that they cannot and should not be like their mothers if they want to be "real men." Likewise, mothers are pressured to disconnect from their sons in the name of being "good mothers." Women raise sons, know them, and love them, yet the forces marshaled to separate sons from their mothers are enormous. The psychological establishment sanctions and encourages this separation and emphasizes the importance of boys' autonomy, independence, self-sufficiency, and disconnection (both literal and symbolic) from their mothers. Thus, sons are taught to abandon women starting with the very first woman in their lives, the woman they are supposed to love best: their mother. Gilligan's early pioneering research with young girls revealed their capacity to comprehend the world of human relationships and responsibilities with a remarkable degree of acuity, sensitivity, and outspokenness (Gilligan, 1982). This research begged the question, if girls could read the relational world so astutely, couldn't boys too? Gilligan asserted that the work of bringing boys back into connection with themselves, with their mothers, and with other women, boys, and men is essential for democracy and for fostering the psychological qualities necessary for citizenship. She pointed out that there is a fundamental tension between democracy and patriarchy: democracy requires love, partnership, and having a voice, while patriarchy relies upon disconnections and silences. The initiation into patriarchy for boys and girls requires a sacrifice of relationship with parts of themselves and with others, and compromises possibilities for full and genuine connections.

To reassert loving and democratic relationships between men and women and to subvert the patriarchal order that promotes the rifts within and between us, Gilligan maintained that men and women must join together as allies. That is, men and women together must support boys' (and girls') healthy resistance to pressures to conform to destructive societal norms. For boys, these pressures to conform to hegemonic masculinity diminish the capacities that are essential for navigating the human world: emotional vulnerability, connectedness, and compassion.

Wilson observed that it is as if boys in our culture are forced to dissociate, to cut off their heads from their hearts and bring only parts of themselves into their relationships and into the world. Gilligan added that a democracy cannot thrive when it comprises a mass of dissociated people unable to bring themselves into authentic relationship with themselves and one another. Working together as allies, women and girls and men and boys face the task of finding a way to allow boys, and all people, to bring their whole, undiminished, uncompromised selves into the world of relationships.

Obstacles to Healthy Development for Boys

Michael Kimmel, professor of Sociology at the State University of New York at Stonybrook and spokesperson for the National Organization for Men Against Sexism, outlined two types of obstacles to healthy development for boys: (1) obstacles that are said to be in the way of boys; and (2) obstacles that really are in their way. An examination of these obstacles—both perceived and real—lends insight into the current sociopolitical climate within which masculinities both shape and are shaped by public discourse and lived experience.

Perceived Obstacles

Within the context of a social landscape marked by the increasing participation of women in public domains, there has been a conservative backlash against the feminist movement. As exemplified by Christina Hoff Sommers' (2000) book, *The War Against Boys: How Misguided Feminism Is Harming Our Young Men*, political pundits and psychologists have put forth the notion that boys need to be "rescued" from feminists. It is against this backdrop that these myths, or the perceived obstacles to healthy development for boys, have emerged.

Myth 1: (All) Boys Are in Trouble

There is indeed evidence that boys are in trouble. As measured by many quality of life indicators, boys lag far behind their female peers in various emotional, educational, and behavioral domains. Boys are, for example, more likely to be diagnosed with ADD, more likely to drop out of school, and more likely to be victims and perpetrators of violent crimes than are their female peers. However, alarmist headlines like one that recently appeared in the *New York Times*, "On Campus, Men Are Vanishing," do not tell the whole story. In actuality, not all men are vanishing from college campuses. Only some are, and typically they are men of color and men of low socioeconomic backgrounds. white men—especially those of the middle class—continue to thrive in many academic and social contexts. The truth of the matter is institutional racism and classism constrain possibilities for some men and expand possibilities for others. Boys' lives play out differently along differing racial, cultural, and social trajectories.

Myth 2: Schools Feminize and Pathologize Boys

Another of the obstacles boys are said to face is their feminization in schools. Schools are accused of enforcing an expectation of a "feminine" docile conformity to obedience in, for example, the insistence that boys sit still, take naps, or speak quietly. This is construed as the pathologizing of a naturally rambunctious boyhood and the promulgation of the message that boyhood is defective. However, as Kimmel pointed out, this position exaggerates differences between boys and girls which are often less notable than differences among boys and girls, and misses the point that interventions designed to benefit girls (for example, attention to new learning styles) may also be beneficial to boys. Feminist efforts to improve opportunities and access to resources ought not to be conceptualized in terms of a zero-sum game such that a gain for girls is considered a loss for boys.

Real Obstacles

The myths surrounding boys' development often obscure the obstacles with which we must wrestle to promote healthy development in boys.

Traditional Ideology of Masculinity

Kimmel noted that the traditional ideology of masculinity is the chief obstacle to healthy development in boys. This masculinity ideology, described as a "cultural myth" by Joseph Pleck (1981) and named the "boy code" by William S. Pollack (1998), represents the values of European American culture and shames young men toward impossible extremes of separation, emotional invulnerability, toughness, and stoicism:

> [T]he middle-class, white, heterosexual masculinity is used as the marker against which other masculinities are measured, and by which standard they may be found wanting. What is normative (prescribed) becomes translated into what is normal. In this way, heterosexual men maintain their status by the oppression of gay men; middle-aged men can maintain their dominance over older and younger men; upper-class men can exploit working-class men; and white men can enjoy privileges at the expense of men of color. (Kimmel and Messner, 1995, 2)

The Invisibility of Gender

One of the most insidious characteristics of the traditional ideology of masculinity is its invisibility to men and boys. Men are treated as if they have no gender, much in the same way that white people are treated as if they have no race. Kimmel recalled a conversation between two female colleagues—one white and one black—in which the former stated that when she looks in the mirror she sees a "woman." The latter, on the other hand, stated that the image she sees reflected is that of a "black woman." Thus for the white woman, race was invisible, while for the woman of color, it was visible and unforgettable. Gender—a mechanism, like race, that both assigns and denies privilege—functions much the same way: men are often considered "genderless" and gender has become a code word for female in this culture. Privilege, Kimmel maintained, keeps privilege invisible. Gender must be made visible to boys, as it is as central an experience for males as it is for females. However, the chief impediment to making gender visible to boys is the unfounded fear that gender equality will result in some kind of a loss to boys and men.

Money

The distribution of public funding reveals a great deal about the systemic and institutionalized perpetuation of masculinity. Kimmel maintained that there is a dearth of public funding for school bond issues (for example, teacher training and support, after-school programs, et cetera) that could support and enhance boys' development by the implementation of new programs, policies, and procedures. Meanwhile, Kimmel noted that large amounts of public funds are being directed to initiatives like the erection of sports complexes, in what he referred to as a "masculinization" of public funding.

Silence

Another set of obstacles centers around the levels of silence and ways in which boys shut down in front of other boys. Boys often assume voices of posturing, posing, false bravado, and impenetrability. When constrained by the traditional ideology of masculinity, other languages—those of compassion, emotional openness, and vulnerability—often are unavailable to them.

Homophobia

Kimmel asserted that the cornerstone of traditional masculinity is homophobia, or the fear of being thought gay. The single most common put-down among boys and men is, "That's gay." Boys police other boys out of their own fear of being seen as "weak," a "sissy," a "faggot," not being "man enough." It is of special note that these insults are not about sexuality per se, but rather about masculinity. Calling a man "gay" is, above all else, an affront to his manhood; as such, homophobia is one of the most imprisoning aspects of the boy code.

Obstacles versus Optimism

Despite the many obstacles, both real and imagined, that boys face in the development of healthy manhood, there are reasons for great optimism. Kimmel cited the rise in students' cross-sex friendships as a hopeful sign of increased understanding among boys and girls. In addition, as feminist women continue to promote the development of healthy boys, there is a greater contribution to the development of a healthy society for men and women alike. Feminism has helped women become more confident, strong-minded, and successful, thus gaining greater access to traits traditionally considered "masculine." By the same token, Kimmel challenged symposium participants to help boys gain greater access to their innate capacities for sensitivity, connectedness, and emotionality, traits that traditionally have been deemed "feminine." Encouraging the experience and expression of the full range of human emotional and behavioral capacity equally offers boys and girls the freedom to be whole.

Building Resistance

Janie Ward, associate professor of Education and Human Services at Simmons College, and project director for the Alliance on Gender, Culture, and School Practice at the Harvard Graduate School of Education, spoke of the need to develop a framework for resistance and resilience. Ward defined resistance as a process by which adults engage with young people to help them figure out how to oppose others' ideas about who they are and should be in this culture. Resistance, Ward noted, is an essential inoculation in a toxic social environment. Drawing from her research with African American youth, but making links to all youth, Ward emphasized the importance of helping children learn to recognize—and oppose—the various "isms" and phobias that contour their lives. She stressed the importance of breaking the silence around racism, sexism, classism, and homophobia, and of speaking the unspeakable in homes, schools, and after-school programs. Ward implicated all adults—parents, teachers, counselors, and friends, among others—in this process, and put forth a model of resistance building that is grounded in four basic but powerful injunctions: read it, name it, oppose it, and replace it.

Read It

The first step of the model is to be aware about what is going on and to talk about it. We must read the relational world in which we live and teach the children in our lives to do the same.

Name It

The next step in the model bids us to find a vocabulary for the cultural experiences and messages that shape and often succeed in limiting us. Ward reminded us of how vulnerable children are to taking in cultural messages about who they are. For African American boys, these messages often include the iconic images of the "gangsta" or pimp, and almost always include the cultural command to assume a "cool pose." The cultural obsession with sex and materialism, our substandard school systems, the prison industrial complex, the onslaught of stereotypical media images, the prevailing silence, disinterest, and dishonesty about race—all shape the lives of African American children and their nonblack counterparts. Adults play an important role in helping children interpret the social world and allowing them to imagine, as Ward described it, "a sense of self greater than anyone's disbelief." Children without a critical cultural analysis are vulnerable children, as they are at risk of viewing themselves through the often distorted lenses of others.

Oppose It

Opposing, or resisting, the cultural strictures that structure our lives is the third step of the model. Not all resistance is healthy resistance, however, and Ward distinguished between two modes: (1) resistance for survival; and (2) resistance for liberation.

Resistance for Survival is a short-term strategy described as an attempt to put together what others have tried to take apart. Resistance for survival can be seen in the "tough guy" stances assumed by some African American boys. These boys—the ones no one can handle, the ones who are always being sent to the principal—are engaging in a mode of resistance designed to protect a fragile sense of self rather than affirm a sturdy sense of self.

Resistance for Liberation, in contrast, is self-affirming. Within this framework, children come to understand that they themselves are not flawed; it is the society that demeans and devalues them that is flawed. Resistance for liberation sets the stage for a liberatory masculinity that, in turn, leads to freedom from gender constraints. It is a resistance strategy that is designed to affirm, rather than to protect, the self.

Replace It

Adults are charged with helping boys replace the myths of masculinity with the truths of their lived experience. Adults may share their knowledge of resistance, be models of resistance themselves, and welcome boys into the community of resistance. Adults may create safe spaces where boys can build and sustain healthy relationships, challenge homophobic behavior, learn media literacy, and feel invited to be their whole, full selves. Strategies of resistance for survival must be replaced with those of resistance for liberation to achieve a liberatory humanity.

New Possibilities for Boyhood and Manhood

In American society-at-large, in the media, and in the social science research, new possibilities for boyhood and manhood are being envisioned, and masculinity is being rethought.

Rethinking American Masculinity

Stand on your own two feet. Be a little man. Be a big boy. Big boys don't cry. Don't be a mamma's boy. Don't act like a sissy. Don't act like a fag.

These all-too-common admonitions give voice to the central messages of the "boy code" that defines and dictates American masculinity. William S. Pollack, director of the Centers for Men and Young Men, director of Continuing Education at McLean Hospital, and assistant clinical professor (Psychology) in the Department of Psychiatry at Harvard Medical School, reminded symposium participants that achieving this masculinity is an impossible task for boys. The values of the dominant European American culture—or "boy culture"—emphasize toughness, stoicism, and violence, and, at the same time, shame boys against emotional vulnerability and relational interdependence. In a process Pollack called "gender straightjacketing," boys become disconnected from their own feelings and from their normative characteristics of vulnerability and need for connection. Anger, which is a precursor to violence, often is the only emotion that boys are allowed to express.

Pollack suggested that behind the anger so often expressed by boys is the stifled genuine voice of the struggle for connection. Boys respond to the culturally enforced code of silence—the "boy code"—which demands that they hide their vulnerability at all costs and avoid the shame associated with it. The sadness, vulnerability, fear, isolation, and despair boys often may feel remains hidden and hard to detect by parents, teachers, and mental health workers, and boys' yearnings for love and affection often are repressed. The toughness and "cool pose" so often assumed by boys are really emotional masks of bravado. Pollack argued that the more we sustain healthy vulnerability in young males, the healthier they will become. He spoke of the need to promote new models of manhood that are connection-based and that allow boys to resist the violence and posturing that have been considered the traditional hallmarks of masculinity. Honoring rather than disavowing healthy vulnerability in boys will lead to a new manhood in America.

Kevin Powell, a poet, journalist, essayist, public speaker, hip-hop historian, political activist, and author, also discussed the necessity of, and some of the challenges inherent in, redefining the American male. On a community level, Powell spoke of the "father void" in the black community where so many sons are being raised by mothers alone. He related his own childhood struggle with self-definition as a boy in the confounding shadow of an absent father: "I knew I didn't want to be like my father, but I didn't know how I was supposed to be." On an institutional level, Powell spoke of the media as one of the foremost shapers of American masculinity, and noted that movies that are particularly appealing to African American youth often depict a florid, glorified violence and feature the message that young black boys must "man up." On a global level, Powell stated that the current sociopolitical atmosphere of war underscores the false notion that violence is the solution to all conflict and reinforces a concept of manhood as defined by the "pistol or the penis."

Drawing from his personal history, Powell charted his own developmental process of definition and redefinition as an American male. He described a childhood induction into the patriarchal order that offered a limited and distorted range of possibilities for boys and men, and for African American males in particular. Powell relayed how, by the time he entered college, his identity as a man was defined by violence, lashing out, flexing, posturing, control, and a sense of superiority over women. When this violence of inner experience dangerously erupted into outward acts of aggression against women, Powell sought help. He spoke about his personal process of redefining himself as a man, which included learning how to listen to women, being honest with his own investment in patriarchy, and writing about his internalized sexism and misogyny (Powell, 1992, 2000). Integral to this process of change was Powell's relationship with a counselor. Within the context of this relationship, Powell had access to a safe space in which he could speak openly, for the first time in his life, with an older, trusted man about his feelings.

Powell asserted that for a "radical revolution of values" to take place, boys and men must create a new paradigm and a new language. Powell reminded symposium participants that there is no such thing as a universal male experience, and claimed that we cannot be silent about the ways in which culture, class, sexuality, and gender affect our lives. Echoing Ward's mandate, Powell emphasized the need for boys and men to learn how to read, name, oppose, and replace the "isms" that shape all our lives. Counselors, mentors, teachers, parents, and other adults have an important role in this transformative process. The work of rethinking American masculinities, challenging and redefining the "boy code," and creating a liberatory masculinity requires an enduring effort on the part of men and women alike.

Rethinking Masculinity in the Media

The media is profoundly implicated both in reinforcing and redressing the boy code; indeed, it is one of the primary pedagogical forces of our time. Jackson Katz, in his educational film, *Tough Guise Teaching Guide: Violence, Media, and the Crisis in Masculinity* (1999), vividly described the media's role in perpetuating hegemonic masculinity. Katz revealed how mainstream media images—from sports, television, Hollywood films, and music videos—help to promote violent masculinity as a cultural norm. He showed that media images of manhood play a pivotal role in making, shaping, and maintaining specific attitudes about manhood.

Patti Miller, director of the Children & the Media Program at Children Now, a research and action organization based in New York and California, further described the powerfully influential force of media on child and adolescent development. She reminded us that children, especially boys, are active users of entertainment and sports media, and spend hours watching television and playing video and computer games. Miller reported that two-thirds of American children have television sets in their bedrooms and spend an average of six hours per day watching them. Children spend more time in front of the television than being read to; in fact, they spend more time with media than they do with any other single activity (Children Now, 1999a, 1999b, 1999c).

Miller highlighted key findings from a recent review of the research on media's messages about masculinity and its impact on boys:

Violence. Media images of violence are pervasive. Violence is used to solve problems and achieve goals; it is depicted as justified, harmless, and without consequence. Male characters typically are portrayed as violent and angry.

Martial metaphors. Sports action is often described in military language. The playing field becomes a battlefield, and sports commentators tend to use terms like "attack," "leave them hurt," "battle lines are drawn," "fighting," and "taking aim" to describe the action.

Vulnerability and emotions. On television, men seldom cry. When they do, it is in isolation. Furthermore, men rarely are presented or perceived as sensitive.

Identity roles. In primetime television, men are depicted as police officers, lawyers, business owners, and other professionals. They are associated with the working world (exterior spaces), as opposed to the world of the home (interior spaces). Men are defined by their careers, whereas women are defined by their relationships.

Homophobia. Gay men are rarely, although increasingly, seen in noncomedic primetime roles. The message is sent that gay men are fodder for jokes and are not to be taken seriously.

These findings beg many questions about the media's role in shaping gendered behavior: Why is there so much gratuitous, glamorized violence on television? Can men be shown to express a full range of emotions without being shamed? Is comedy used to enforce gender-role stereotypes and homophobia? How might the media's stereotyped portrayals be redressed? How might other, healthier, possibilities for boyhood and manhood be portrayed?

Marjorie Cohn of Nickelodeon, in her discussion of her network's programming, offered some partial responses to these questions. She spoke of the importance of showing many diverse portrayals of boys: boys sharing emotions, boys breaking stereotypes, boys engaging in cross-ethnic friendships. She also acknowledged the need for media programming to dig even deeper and produce shows conceived by people other than the white, middle-class males who currently create the majority of television shows.

There is an undeniable need to provide boys and girls the conceptual and practical tools for reading media images critically. Young people must be supported in their quest to make sense of the apparent contradictions between the truth of their realities and the media's account of the truth. Teaching young people to engage in a critical analysis of harmful media images diminishes these images' insidious capacity to shape and distort self- and others' perceptions. The media has the power to portray more authentic versions of the male experience. It should be held accountable for the potentially damaging images it projects and be encouraged "to provide boys a fuller, more complete picture of the men they can become" (Children Now, 1999a, 21).

Rethinking Masculinity in Social Science Research

There has been a recent resurgence of interest in the empirical study of boys and men that has come in the wake of the pathbreaking feminist research on girls and women. Feminist psychology has generated important knowledge about gender, innovative research methodologies, and new understandings about human development, which in turn have begun to impact the ways in which boys and men are studied (Way and Chu, 2004). Ritch Savin-Williams, professor of Human Development at Cornell University, is among the cohort of scholars who are rethinking masculinities in the social science research. His work on gay teens offers new insights into the nature and significance of same-sex sexuality for boys.

In his discussion of the "new" gay teen, Savin-Williams (2005) introduced the concepts of "postgay" and "gayishness." These terms capture the contemporary reality of teenagers, who increasingly are engaged in the renegotiation and redefinition of their sexualities to such an extent that sexual identity labels—like "gay"—are rendered meaningless. The teenagers in Savin-Williams's study used labels like "pansexual," "heteroflexible," "queer boi," "trisexual," "trannyboy," "omnisexual," "boidyke," and "multisexual," among others, to describe themselves. With terms like these, teenagers literally are reshaping the language to reflect more accurately their reality and to acknowledge and name both their gender and their sexuality.

Societal constructions of gay teens have been changing and evolving right alongside teens' individual and collective self-definitions. With the invention of the concept of "gay adolescence" in the 1970s came the prevailing message that gay youth were in deep trouble: suicidal, despairing, drug addicted, and lost. Savin-Williams named the consequences of these negative constructions (which, he noted, were based on flawed research), describing how they pathologize gay youth and feed the agenda of the religious right. Savin-Williams also noted that these negative constructions divert attention from those teens who are most at risk, ignoring the fact that the majority of gay youth are resilient and, in fact, really quite ordinary.

Today, researchers and practitioners understand that young people have a range of same-sex attractions and that the spectrum of sexualities is broad. Many gay teens are resilient and proud, and say they feel supported by the vast majority of their peers. Despite the persistent presence of regressive social forces and institutions, trends and images in the popular culture signal the assumption of (rather than entreaties for) the acceptance of homosexuality. Savin-Williams anticipated a future in which gay adolescents will not be considered unusual. Instead, they—and the fluid spectrum of sexuality they possess—will be recognized as nothing more or less than ordinary.

Breaking the Link between Masculinity and Violence

The film *Tough Guise* illustrated how manhood, as it is culturally constructed in our society, is related to power, control, and violence. Instances of extreme violence (like the school shootings at Columbine and elsewhere) are cast into relief against a backdrop of normative, everyday violence such as that enacted on athletic playing fields, in international public policy, in interpersonal relationships, and by cultural heroes such as Arnold Schwarzenegger's *Terminator* or Sylvester Stallone's *Rocky Balboa*. In social, political, and economic institutions, regular men are seen acting violently. Violent masculinity is culturally normative, rather than unusual, unexpected, or intolerable; it is the roadmap by which boys become men. Violence relies on a lack of emotional connection, a distortion of relationship. The effects of men's violence are felt not only by women but also by children and by other men with less power.

Katz asserted that the key step in reducing violence is to change definitions of manhood and develop a new language of accountability and connection. This is an undertaking in which men must play a central part. Kimmel pointed out that the very phrase "violence against women" is grammatically incorrect. It contains an object (women), but no subject (men). The rampant use of the passive voice when talking about crimes against women serves to shift the focus from male perpetrators onto female victims and survivors.

Katz noted that traditionally, strategies to prevent men's violence against women have not been "preventive"; instead, they have been cautionary or accusatory injunctions aimed at women. Women have been admonished to proverbially "take back the night" by, for example, being careful to guard their drinks to avoid being slipped the "date rape drug," or by being counseled to leave abusive relationships, or by being warned not to dress provocatively. Rarely have men been charged with the directive to "give back the night," or to join with other men, or with women, to create safety in communities and interpersonal relationships. Focusing on women's role in men's violence against women with questions such as "What was she wearing?" or "Did she try to fight him off?" has served to divert attention from the more appropriate, more politically charged questions such as, "Why are men doing this to women?" and "How can we make them stop?" Ending men's violence against women has been left up to women for far too long; the time has come for men to participate in the struggle for social change. Katz was among several panelists in the symposium who described their practical, community-based efforts to work with men to redefine manhood and break the link between masculinity and violence.

Mentors in Violence Prevention Program

In addition to creating *Tough Guise*, Jackson Katz, a former all-star football player, founded, in 1993, the Mentors in Violence Prevention Program (MVP). Based at Northeastern University's Center for the Study of Sport in Society, MVP encourages men to engage actively in the prevention of men's violence against women. Through MVP, what traditionally have been seen as "women's issues"—rape, sexual harassment, and domestic violence—become men's issues as well. MVP aims to help athletes at all levels—high school, college, and professional—to develop an awareness that does not equate strength in men with dominance over women.

Don McPherson, a former National Football League (NFL) quarterback, was formerly national director of the MVP Program. McPherson, who now serves as the executive director of the Sports Leadership Institute at Adelphi University, spoke of men's responsibility to work proactively to end violence against women. He stated that violence against women stems from men's attitudes about women and from the rigid, restrictive policing of masculinity that pits men against women. McPherson noted that in our society we do not raise boys to be men, rather we raise boys not to be women.

Boys are raised on a diet of negative, misogynistic injunctions— "Don't act like a girl," "Don't be a pussy"—that are based on the degradation of girls and women. The command to "be a man" becomes a code for an emotional shutdown that boys learn almost as soon as they learn to cry. McPherson asked, "Do we make our boys stronger by making them tough?" The answer is no. "Being a man" debilitates men and undermines their empathic capacities; it silences them into complicity. Men are made weaker by the demands that they be strong.

Strategies employed by MVP include encouraging men to break the complicit silence that quietly but powerfully condones the violence of men against women. MVP teaches young men how to be emotionally connected, empowered bystanders who are able to confront abusive peers and, moreover, offers them specific scripts and strategies by which to do so. The program creates a safe physical and emotional space for boys and men to work together to model and enact transformative ways of being in relationship to one another and to girls and women.

Men Stopping Violence

Sulaiman Nuriddin is the team manager of Men's Intervention Programs at Men Stopping Violence, an organization that works locally and nationally to dismantle sexist belief systems, social structures, and institutional practices that oppress women and children and dehumanize men themselves. Challenging male domination and patriarchy, Nuriddin and his colleagues work directly with men who are violent, many who have been court-ordered to attend the center's class-based intervention programs, and others who have chosen to participate of their own accord. Nuriddin described three prevention and intervention strategies that he utilizes to challenge sexist structures, foster an understanding of male privilege, prevent violence, and promote change among these men:

Men modeling behaviors for boys. An integral part of the program is men demonstrating for boys ways to avoid abusive behaviors. Nuriddin and his colleagues noticed that the men who entered their program often had sons at home who had borne witness to the abuse they had committed. These boys were invited to the intervention sessions so they could now bear witness to their fathers engaging with, challenging, caring for, supporting, and learning from other men. Comprising group meetings, journal writing, and mentoring, the program is designed to transform beliefs about what it means to be male.

Community engagement. For these antiviolence efforts to be successful on the individual level, they must involve the community as well. Men and boys leave the special space created by the program and return to their neighborhoods, where their new ideologies, behaviors, and commitments are not reinforced and are even undermined.

Nuriddin and his colleagues saw the need to create a critical mass of men trained to do this work in their communities. Just as fathers were encouraged to invite their sons to participate in the program, so now are boys encouraged to invite their friends. The men and boys challenge themselves, one another, and their communities to stop men's violence against women.

White Ribbon Campaign. Another example of a program that provides men with ideological and practical tools that aid in their personal transformation, challenge the established gender order, and promote human rights is the Canadian-based White Ribbon Campaign founded by Michael Kaufman. The White Ribbon Campaign is the largest effort in the world of men working to end violence against women. As a part of this effort, men around the world are urged to wear a white ribbon each year for one to two weeks, starting November 25, the international day for the eradication of violence against women. During this time, men are encouraged to speak out about the problem of violence in their homes, workplaces, places of worship, and communities. Wearing a white ribbon serves as a pledge to never commit, condone, or remain silent about men's violence against women.

The White Ribbon Campaign is a nonpartisan, decentralized, grassroots campaign that is alive in many different countries. With its public education and action kit, it offers a framework for men to create spaces for discussion and action in their own communities. This campaign, like the Mentors in Violence Prevention and Men Stopping Violence programs, combats the overwhelming social silence and cultural complicity surrounding men's violence against women.

These programs encourage a kind of reflection, self-interrogation, and discussion that leads to personal and collective transformation and action by men. By turning men into public educators and active witnesses who resist committing or condoning men's violence against women and who accept accountability for their actions, these programs contribute to the safety and human rights of all people, and help create a more just and less violent world.

Following the presentation of the Mentors in Violence Prevention Program, Men Stopping Violence program, and White Ribbon Campaign, panelists and audience members engaged in a conversation about the broad spectrum of men's violence against women. It was acknowledged that physical violence exists along a continuum that includes the emotional violence of sexist joking, verbal sexual harassment on the street, and other domineering, demeaning forms of thought and behavior. The conversation also turned to the importance of addressing men's violence not only on the individual, local, institutional levels, but also on the international level. Women's bodies at home and abroad often are seen as little more than commodities that can be bought or sold for the comfort and pleasure of men (as is evidenced by the U.S. military's unwritten recruitment and reward strategy that involves the exploitation of female bodies overseas). There is a need to change the social norms that create the context for men's violence within and beyond our borders. The issue of men's violence is not a "woman's issue"; it is a legitimate human concern that shapes society, affects interpersonal relationships, and drives public and international policy.

Remaining Challenges

Pedro Noguera, professor of Sociology at the Steinhardt School of Education at New York University, closed the symposium with a discussion of the remaining challenges faced by those of us committed to the work of supporting boys' resilience. Casting the self-described voice of pessimism, Noguera raised his reservations about the possibility of undoing patriarchy. It is, indeed, difficult to conceive of a movement on the part of those with privilege (in this instance, men) to relinquish that privilege and redistribute their power. Noguera noted that there are great limits to this work because it asks those who are benefiting to make significant, and perhaps ultimately insufferable, sacrifices.

The symposium, with its primary focus on change and growth at the individual and community levels, paid less attention to the ways in which institutional structures of power reinforce patriarchy. Militarism, politics, and capitalism are all structures that promote and are fueled by the dominance of some people over others. We cannot focus on the individual without acknowledging the contexts and circumstances that shape their behaviors and produce or preclude their possibilities. The answers to the problems inherent in hegemonic masculinity lie beyond simply expanding the range of emotions available to individual men. Prescriptions for addressing and redressing these issues cannot overlook the structural imbalance of power which privileges middle-class white men over working-class white men and men of color, differently marks their access to resources, and variously shapes their opportunities and life chances.

Noguera reminded us that this is not a simple story of victims and victimizers. He spoke of groups of men who are themselves victims of a certain sort—of racism and/or of poverty. He noted that it is hard to convince certain men that they are powerful when they do not have jobs, when they cannot provide for their families, when they are subject to daily discrimination. Noguera asked, "What does it mean to be asked to give up power when you feel powerless?" Men's relationship to gender-based domination must be located within the context of the social injustices that structure society for women and men alike.

Bringing African American males into focus, Noguera spoke of the multitude of social problems they face, including an unemployment rate of 50 percent, a declining life expectancy, and appallingly high rates of incarceration. In schools, African American boys struggle with lagging grade point averages, high expulsion rates, and poor graduation rates. Further, research shows that African American boys become increasingly disidentified with academic achievement as they move from eighth to twelfth grade, such that by the time they finish high school, there is no longer a relationship between these boys' self-esteem and their grade point averages (Osborne, 1997).

This research finding suggests that these boys, so often disidentified with school, are seeking affirmation elsewhere. In a society such as ours, in which there is little ego affirmation for men of color, a traditional masculinity ideology may be the last thing to hold onto.

Noguera conceded that masculinity for all men, regardless of race or class, must come to stand for attributes that affirm rather than diminish humanity. However, he reminded symposium participants that it is a complicated undertaking to endeavor to remove the "tough guise" from boys who are living on tough streets; boys living in hostile environments will be victimized if they show weakness. Noguera said we must respond to the reality of the various worlds in which boys live and develop their assorted—and unequal—masculinities. Among the societal institutions that produce culture—media, family, places of worship, and schools—schools are the sites of socialization where we have the greatest ability to influence. Noguera reminded participants that gender socialization within schools is an essential part of the hidden curriculum. A new set of curricula must be developed—and in this area there is a dearth of theory and research to guide us—to promote an affirmative and liberatory masculinity for boys of all races and classes.

Conclusion

This symposium provided an important space within which to address critical questions about masculinity. However, many difficult questions remain only partially or not at all explored: How do we shift the focus from changing the violent behavior of individual men to mobilizing men to challenge systems of gender violence and related structures of oppression in the United States and abroad? How do we include the voices and experiences of men who are not white and not black? That is, how do the experiences of Asian boys and men, or Native American boys and men, inform and expand this conversation? Where are the examples of men and boys, in their everyday lives, being resilient and resisting a capitulation to hegemonic masculinity? How can we more effectively involve mothers and women in this work?

We must create changes in society on the cultural, structural, and individual levels so as to enlarge the space for boys to be human beings; we must substitute a masculinity that is synonymous with invulnerability and indifference with one that is constituted by openness and compassion. As Jackson Katz and J. Earp (1999, 5) propose in their *Tough Guise Teaching Guide*:

> In the final analysis, what's required is a full-scale transformation in how we imagine, define, and model masculinity—a personal and institutional revisioning of manhood that specifically and forcefully affirms courage as something far more noble than simply possessing physical prowess and power. This means nothing less than holding to a vision of masculinity that is entirely at odds with senseless violence, bullying and posturing, and entirely in keeping with grace, compassion and the guts to stay loyal to what's right.

References

Children Now (1999a). *Boys to men entertainment media: Messages about masculinity*. Oakland, CA: Children Now.

———. (1999b). *Boys to men sports media: Messages about masculinity*. Oakland, CA: Children Now.

———. (1999c). *Fair play? Violence, gender, and race in video games*. Oakland, CA: Children Now.

Gilligan, C. (1982). *In a different voice: Psychological theory and women's development*. Cambridge, MA: Harvard University Press.

———. (2003). *The Birth of Pleasure*. New York: Vintage Books.

Katz, J., and Earp, J. (1999) *Tough guise teaching guide: Violence, media, and the crisis in masculinity* (video). Northampton, MA: Media Education Foundation.

Kimmel, M. S., and Messner, M. S. (Eds.) (1995). *Men's lives*, 3rd edition. Boston: Allyn & Bacon.

Osborne, J. W. (1997). Race and academic disidentification. *Journal of Educational Psychology* 89: 728–736.
Pleck, J. (1981). *The myth of masculinity*. Cambridge: MIT Press.
Pollack, W. S. (1998). *Real boys: Rescuing our sons from the myths of boyhood*. New York: Random House.
Powell, K. (1992). The sexist in me. *Essence Magazine*, September, 17.
———. (2000). Confessions of a recovering misogynist. *Ms. Magazine*, January, 21–23.
Savin-Williams, R. (2005). *The "new" gay teen: Post-gay and gayishness among contemporary teenagers*. Cambridge, MA: Harvard University Press.
Sommers, C. H. (2000). *The war against boys: How misguided feminism is harming our young men*. New York: Simon & Schuster.
Way, N., and Chu, J. (Eds.) (2004). *Adolescent boys: Exploring diverse cultures of boyhood*. New York: New York University Press.

Part II

Collaborations and Program Assessment

Chapter Eight

Power and Possibilities: Collaborative Fund for Youth-Led Social Change

Ami Nagle, Marisha Wignaraja, P. Catlin Fullwood, and Margaret Hempel

Exposing the myths of community revitalization. Making schools safer. Promoting after-school opportunities. Ensuring that workplaces are safe for young workers. Reducing sexual harassment of young women in the juvenile justice system. Raising awareness of the health care needs of Latinas. Fighting for the rights of immigrant workers. Youth are changing the world.

Youth development and youth organizing efforts are demonstrating that youth can make a positive difference in their own lives and the lives of their families and communities.

Partnering with youth in social change efforts, and supporting their growth and development is no easy task. It must be done in ways that help youth build personal and professional skills—and it can only be fully accomplished by programs that address the identities that shape youths' lives: race, class, gender, and sexual orientation. To ignore these factors and their interrelation is to fundamentally misunderstand youth and to neglect a vital opportunity to help them marshal their strengths, overcome barriers, and change the world around them.

Recognizing the power and possibilities of working at the intersection of youth development, youth organizing, and programming that recognizes the importance of multiple youth identities, the Ms. Foundation for Women and its partners launched the Collaborative Fund for Youth-Led Social Change (CFYS).

This exciting initiative works with donor and grantee partners from across the United States to

- Explore a variety of social change and organizing models that are steeped in community, organizational, and political contexts.
- Support intergenerational leadership efforts that promote learning and mentoring between adults and youth, and provide

youth with real leadership opportunities in programs and organizations.
- Build capacity of innovative organizations to undertake program activities and build organizational strength and longevity.
- Develop a learning agenda that ensures that participating organizations and the field learn from this unique initiative.

A key to the success of CFYS is the strong commitment to promoting diversity. This diversity takes many forms—from the wide range of grantee and donor partners to the variety of organizational types and approaches to both youth development and youth organizing.

Supporting this initiative is a framework of equal participation among donors, youth organization staff, and youth leaders. The partners work together to shape initiative activities and information-sharing opportunities. Partners value the opportunity to develop working relationships with a variety of individuals concerned about a holistic approach to youth development and organizing. These relationships ensure that each organization's approaches and practices continue to evolve.

It is out of this partnership that the learning component of CFYS will be enveloped. Based on questions from youth, youth organization staff, and donors, CFYS will engage all partners in a process of identifying, discussing, and documenting promising practices. This learning will inform the future activities of organizations that serve and support youth.

This chapter provides an overview of this initiative; the tenets upon which this work is built; the kinds of work grantee organizations are pursuing; the learning network created by this initiative; and the kinds of questions that donor, youth organization staff, and youth partners hope to address by participating in this fund. Much has changed in the youth fields and programs for youth in the last 30 years.

Reacting to negative stereotypes of youth promoted in the 1980s and 1990s, the youth development field began to advance a positive youth development frame. This frame, and programs built on it, underscore that youth can and do positively impact their own lives and communities (Pittman, 1991).

Over the last ten years, the fields of youth development and youth-led social change (also called youth organizing or youth civic engagement) have begun to recognize the benefits gained by merging strategies (Cahill, 1997). Some in the youth development field have recognized that many adolescents are naturally interested in questioning the social constructs that surround them. However, in working to develop individual skills, youth face societal barriers to their individual development, such as classism, racism, gender discrimination, poor schools, and lack of economic opportunity. At the same time, some practitioners and funders working on youth organizing have realized that to be effective and responsible community organizers, programs needed to address the developmental needs of adolescents. The two arenas naturally complement each other.

Innovative work, including the partnership between the Ford Foundation and the Innovation Center for Community and Youth Development, and local and regional youth organization efforts supported by the Tides and Surdna Foundations, and the Funders Collaborative on Youth Organizing, helped these fields better understand a new approach—an approach that is dedicated to individual development through civic engagement and community improvement through collective action (Innovation Center for Community and Youth Development, 2000).

Youth in the Lead

Youth spearheading efforts to improve their own lives and the health of their communities, workplaces, and schools is just part of the story. Many groups are moving to fully entrust the design and implementation of programs to youth participants. However, if we are serious about youth leading organizations and having a voice not just in the design of the project activities but also in the development of the youth program and the larger organization, we need to further develop intergenerational power sharing models for decision making. Some youth programs are developing innovative approaches to involve young people in organizational decision making, including involving youth on boards, as staff, and in fundraising, program, and strategic planning. New models require adult staff to balance teaching and mentoring with providing youth with the opportunity and training to make and carry out organizational and program decisions. CFYS has found that this also requires a genuine commitment to two-way learning where youth and adults work with each other, learn from each other, and share leadership roles. Building organizational leadership skills and engaging youth in these tasks is not easy, but when the right conditions come together it can be a powerful strategy for positive change (Zeldin et al., 2000).

Youth Development and the Role of Identity

Some in the youth development community have begun to explore how young people's development is shaped by factors including gender, race, and class (Camino, 1995). While gender is listed as an identity in much of this work, little has been done to understand if and how girls and boys programs need different approaches because of differences in life experiences and gender norms. An analysis of existing work showed that the most capable of these programs explored the social construction of gender and invited young women and men to challenge traditional roles, examine gender privilege, and create an even balance of power between them (Mead, 2001). Recent research finds that organizations specifically focusing on and supporting a diversity of youth identities are often the most effective at making youth feel comfortable, helping them assume leadership roles, and raising awareness of the role of identity and discrimination (Roach, Yu, and Lewis-Carp, 2001).

While work in this area is increasing, few efforts concentrate on understanding the connection between gender identity and sexual orientation, and its role in youth development. CFYS incorporates gender identity and sexual orientation into a broader definition of gender that includes not only male and female, but also lesbian, gay, bisexual, transgender, or gender questioning. Even fewer efforts explore how the intersection of gender, race, and class identities affect youth participation and experiences.

Earlier work of the Ms. Foundation's Collaborative Fund for Healthy Girls/Healthy Women contributed to the field's understanding of the benefits of youth development programs that explicitly address the needs of girls and young women. Research from this initiative examines how programs recognize girls' voices and foster their individual and group empowerment. Effective programs created "safe space" for girls, recognized different approaches to leadership, established intergenerational relationships among girls and young women, and promoted opportunities for girls to improve the world around them (Ms. Foundation for Women, 2002). Emerging from this research was a recognition of the different strategies for social change that enable young women to make a difference in their lives and communities.

Creating a better world for girls and young women is about the delicate balance of building and strengthening relationships among girls and young women, and also building and strengthening their relationships to women, boys, men, and the broader society around them. Most young women in youth development programs are in mixed-gender environments. Focusing on what happens with young women as well as young men in these environments will move this partnership and the field closer to the goal of gender, race, and class equity.

Understanding the importance of strong young women and young men to the lives of all youth and the future of communities, the Ms. Foundation moved to include boys in public awareness and programming. Most notably, the Ms. Foundation shifted its successful ten-year-old Take Our Daughters To Work® program, which focused on making girls visible, valued, and heard, to Take Our Daughters And Sons To Work® Day. The new program encourages both girls and boys to share their expectations for the future and think about how they can participate fully in family, work, and community. The program also challenges workplaces to consider policies that will help their female and male employees better integrate these multiple demands.

A combination of these forces led Collaborative Fund partners to realize the power and possibilities of the next phases of work—work that is positioned at the intersection of youth development, youth-led social change, and gender.

Youth Development and Youth Organizing: A Spectrum of Social Change

Involving youth in social change requires a shift from unquestioning acceptance of the way things are to developing a strategy that engages communities and institutions to address injustice at a systems level. Such efforts are described and understood in many ways. To help define this work, Collaborative Fund for Youth-Led Social Change (CFYS) partners recognized and developed a spectrum that represents the forms of social change that youth and adults can engage in within different contexts and at different points in time. This working definition will be refined as the initiative unfolds (figure 8.1).

This spectrum is not linear; rather, it is opportunistic, enabling participants and programs to engage in a variety of strategies shaped by the context of the community, background of the effort, needs of youth participants, and the social change desired.

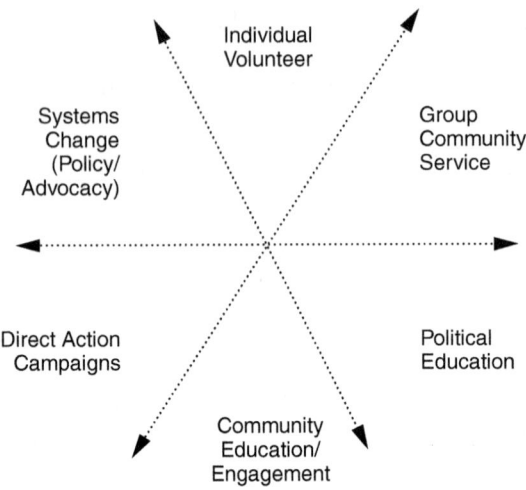

Figure 8.1 Spectrum of Social Change.

To assist groups in evaluating their own readiness to take on youth development and youth organizing work, CFYS partners developed a series of questions used on site visits. Some questions included:

- Who sets the social change agenda and how is it set?
- Does the youth-led social change action have potential for long-term systems change?
- Does the youth-led social change action address issues beyond those normally seen as "youth issues"?
- How does the organization incorporate issues of gender, race, class, and age into its social change work?

Participants in CFYS come at their social change work from a variety of perspectives:

Khmer Girls in Action. Working with young women of Southeast Asian descent, Khmer Girls in Action (KGA) builds young women's capacity to assess how their physical, emotional, and mental well-being is influenced by political, social, cultural, and economic factors, and to develop action campaigns to address community issues. A major concern for KGA members is the devastating impact of anti-immigrant policies on their community. Because of new repatriation agreements, over 1,500 Cambodians who have lived in the U.S. for most of their lives (many of whom have children who are U.S. citizens and who may be the sole breadwinner for their family) are being deported. KGA identified this as a crucial issue and, among other activities, mobilized 300 youth and community members in downtown Los Angeles to advocate for immigrant rights.

Appalachian Women's Leadership Project. Beginning in the summer of 2002, the Girls Resiliency Program (GRP) began alerting people across rural Lincoln County in West Virginia to the imminent closure and consolidation of four junior/high schools. Youth leaders of the GRP pressured the State Board of Education to move back the date of closure and hold public meetings. Young women designed and made badges, created posters, and prepared speeches voicing their opinions about the school closures. While the State Board of Education created road—blocks to public participation—requiring signing up 24 hours in advance to provide testimony at public hearings, and holding meetings on sudden notice—the youth and community members they had mobilized were successful in voicing their opinions to key decision makers. In addition, GRP has been working with a lawyer and community members from affected school areas to file an injunction against the school consolidation plan.

Girl Scouts of the Milwaukee Area. With the hope of establishing City Action Teens Teams (CATT) throughout Milwaukee County, the Girls Scouts of the Milwaukee Area is helping young women come together, identify issues of concern to the community, and promote individual and community response. For example, this year, one of the CATTs tackled the problem of lack of information about and access to health services in the Latina community. The group developed and implemented two community health fairs, securing major health provider, business, and government partners who provided free health education, awareness, services, and referrals to participants. The CATTs also initiated a Teen Summit that brings girls from across the city together in a safe environment to articulate their concerns, connect to services that can support them, and identify action they can take to become change agents in their communities.

Sisters in Action for Power. At a strategic planning meeting early in 2002 designed to identify a new issue campaign, members at Sisters in Action for Power in Portland began talking about how their neighborhood was changing. Young women discussed the seemingly endless money available for construction, the overnight replacement of established markets and beauty shops with wine shops and art galleries, and forced displacement of families and friends from the neighborhood. At the same time as the term "revitalization" was used to promote and describe these neighborhood changes, young women were frustrated at the lack of resources and attention to "revitalize" public schools and housing. Portland Public Schools announced the sale of 60 acres of land, and a federal HOPEV I grant was given to Portland to demolish the largest public housing complex in the state. Sisters in Action for Power launched their new Land Equity campaign to dispel the myths of revitalization and fight the dismantling of public schools and public housing.

The Ms. Foundation for Women and its donor partners launched the Collaborative Fund for Youth-Led Social Change (CFYS) to further the fields of youth development and youth organizing, explore the role of gender identities and orientations in youth programming, bolster the efforts of innovative youth organizations and programs across the nation, and move forward the learning that will advance work across these areas.

CFYS donor partners represent a spectrum of giving within philanthropy. From women's funds and community foundations, to family, corporate, and independent foundations, to individual donors, these organizations and individuals emphasize a range of giving and interests. Some emphasize national giving, while others concentrate on a state or local community. Their issues include girls' and women's rights, youth development, community and youth engagement, and antipoverty strategies. This rich diversity of donor partners enables CFYS to live up to its mission and better support the grantee partners of the fund.

The mission of CFYS is to create, sustain, and enhance a thriving network of funders and local youth-serving organizations that demonstrate the power and possibility of young women and men to create positive change in their lives and their communities, schools, and workplaces. Together, partners—including funders, youth organization staff, and youth—share models that recognize individual development, social change, and gender identity as integral to the process of building an equitable society.

To reach these goals, CFYS partners identified three overarching fund objectives:

- Support and document innovative social change models that combine the best practices of positive youth development and youth organizing within programs that understand that young women and young men engage and develop differently.
- Strengthen the connection between girls' programming and youth-serving organizations.
- Leverage resources from diverse funders to bolster the youth development and social change efforts of innovative girl-only and mixed-gender programs.

The fund includes three years of grantmaking, technical assistance, and knowledge-sharing opportunities. The strength of CFYS is built on several core tenets:

Partnership development. By equally engaging funders, youth organization staff, and youth participants, CFYS is a unique learning community that works to further the intersection of youth development and youth-led social change. The partnership will develop and promote approaches that recognize the importance of multiple youth identities to youth development, the range of approaches to successful social change, and the importance of accepting a diversity of youth leadership styles.

Grantmaking. Financially supporting organizations at the forefront of this innovative work through multiyear grants is crucial to ensuring that their work continues and programs have the capacity to reflect on key lessons.

Capacity Building and networking. CFYS includes capacity-building and networking activities that engage funders, youth organization staff, and youth partners. For the funders, participating in a collaborative fund and engaging in real dialogue with program partners builds their understanding of the needs of the field and how best to support organizations and their programs. For youth organization staff, networking with peers and funders enables them to explore different organizational and program models that can help reinforce efforts, broaden their experience, and build a common movement. For youth partners, networking with youth leaders and donors from across the country reinforces youth leadership development, exposes them to other approaches to social change, and supports their efforts to assume leadership positions within their organizations.

Learning and dissemination. Developing and supporting the work of cutting-edge efforts is most useful when key program components are carefully documented, lessons are extracted, and information about innovative and effective practice is disseminated to the field and funder communities. CFYS works with all partners to develop strategies to systematically document initiative activities and disseminate findings to key constituents.

The Grantmaking Process

With the mission, objectives, and core tenets in place, CFYS donor partners released a Request for Letters of Intent. In response,

576 youth-serving organizations from across the nation submitted letters. CFYS donor partners then collaboratively solicited proposals, reviewed materials, developed guiding questions and selection criteria, and conducted site visits. Through this intensive process the partners gained clarity about the intent of CFYS, the strengths of applicant organizations, and potential learning opportunities presented by the initiative.

By summer 2002, 12 organizations were chosen because of their proven track record in at least two of the three main issue areas, their commitment to and vision for excelling in all three areas, the organization's ability to contribute to national learning, and readiness to undertake the challenge of taking their work to the next level. In addition, priority was placed on creating regional representation, racial/ethnic balance, and on exploring how gender is addressed within the organization. Each organization received $105,000 over three years.

The Work of the Organizations

The Collaborative Fund for Youth-Led Social Change is a learning community of youth organization staff, youth leaders, and donors. Our grantee partners, representing organizations from across the country, take the lead in exploring the power and possibilities of working at the intersection of youth development, youth-led social change, and programming that recognizes the different and similar developmental needs of young women and young men.

Participating organizations vary widely, providing a rich partnership for developing and learning about strategies, tools, and models. Key differences include

Geography. The organizations represent a wide spectrum of places, with one from the South, four from the West, one from the Northwest, three from the East, and three from the Midwest. They work in both urban and rural areas. They recognize the unique context of their community and adapt their strategies accordingly.

Organizing strategies. These organizations use a variety of tactics from popular education and campaign development to community mobilization and peer-to-peer education.

Strengths. Some come from youth development origins while others grow out of social change movements. All begin with a frame and understanding of young people's strengths and possibilities.

Participant constituencies. Eight organizations work exclusively with young women while the remaining four work with young women and young men. They represent African American, Asian American, Caucasian, Hispanic, multiracial, and immigrant communities.

Organizational structures. Each program has different organizational auspices and histories. Some are independent, youth-focused organizations and some are part of larger organizations. Two groups are affiliates or part of larger national networks.

While some factors are unique to each organization, there are recurrent themes that link them together, including:

Supporting youth in decision-making roles. Each organization is committed to engaging youth in program and institutional decisions. Further, these organizations will work to better understand power sharing between youth and adults in projects and organizations. For example, program and youth partners will explore ways to engage youth in planning, finance and budgeting, fundraising, board and staff development, and evaluation.

Working together to change community conditions that primarily impact youth. Each organization works with youth to identify the issues that impact their lives, the lives of other youth and the communities around them, and craft action strategies to address those conditions.

Building new leaders. Each organization works with youth to develop and lead social change efforts and shape organizations. CFYS partners recognize the diversity of leadership styles and the importance of creating venues where a variety of young people have opportunities to lead. In building new leaders, organizations will explore the balance of youth and adult partners in leadership development.

Exploring the role of gender in programming. Research and practice has helped these organizations understand that "gender matters." In different ways, they address gender privilege and the role of race, class, and sexual identity in society, and work to eliminate societal biases. Gender identity and orientation is broadly defined to include not only male and female, but also lesbian, gay, bisexual, transgender, gender questioning, et cetera.

Developing national learning and organizational capacity. These organizations are motivated to share, learn, and document successful strategies. They build capacity through examining organizational approaches and activities and working to expand and strengthen their work. They are looking to improve their own work and that of organizations across the nation, and learn from current developments in the youth fields. They are committed to active participation in a learning community of youth practitioners and funders.

Gender Identities in Programming

Gender identity matters in youth programming. Understanding how youth think about gender and how it impacts their lives is a crucial component to building strong, successful youth development and social change programs. To reflect how many youth think about these issues, CFYS defines gender broadly to encompass gender identity and sexual orientation, including male, female, lesbian, gay, bisexual, transgender, and gender questioning. This working definition will be refined as the initiative unfolds.

To assist groups in evaluating their own readiness to incorporate gender identity into youth development and youth-led social change work, CFYS developed a series of questions used on site visits. Some questions included:

- Does the program consciously address and support unique gender needs through program space, curriculum, education, training, safety of location, and/or access?
- What are the leadership opportunities available to young women and young men in the organization?
- What are the similarities and differences in the participation of young women and young men?
- How is the youth development and youth-led social change work supportive of a broad range of gender identities?
- What are some of the different approaches the program takes that respond to the unique needs of young women and young men?

Participants in CFYS address gender in a variety of ways, including:

Blocks Together. Focused on Chicago's northwest side, Blocks Together works with a Youth Council of young men and young women to identify concerns, increase youth's skills to advocate for themselves and their community, and develop leadership skills to win tangible changes. Trainings and exercises have helped youth council members discuss gender, race, and class discrimination issues. To address harassment by school security guards, the members of the Youth Council, in coalition with Chicago Youth United, started collecting information and stories from their peers. While the experience with security guards was not uniform across the city or even within an individual school, the youth noted that young women tended to experience sexual harassment whereas young men tended to be victims of physical violence. Both the young men and the young women realized that while the problems they faced were somewhat different, the solutions were the same. They are working to change the hiring and training practices for security guards.

PEARLS for Teen Girls. As a girl-only program in Milwaukee, Wisconsin, PEARLS for Teen Girls employs teen facilitators to create safe space for girls ages 11 to 13, explore barriers, and create a better world for themselves and the women around them. Beginning with individual goal setting exercises, PEARLS and the teen facilitators work with the girls to create group and community goals. Especially focused on how they as young women can help improve the lives of other young women in the community, PEARLS is tackling the issue of increasing access to comprehensive sexuality education in public schools, through popular education strategies, theatrical performances, creating fact sheets, and accumulating signatures to petition the Milwaukee Public Schools to require sexuality education in public schools.

Center for Young Women's Development. Working to develop the skills of young women involved in the juvenile justice and foster care systems, the Center for Young Women's Development (CYWD) in San Francisco provides training and employment opportunities, identifies issues of concern to this vulnerable population, and crafts action strategies. Young women incarcerated in juvenile prisons are often vulnerable to sexual harassment. To address this issue, CYWD leaders have gathered information from incarcerated youth, researched antidiscrimination policies, and pressured juvenile corrections to institute a landmark antidiscrimination policy to protect incarcerated lesbian, gay, bisexual, transgender, and queer youth. Youth participants are now developing and implementing sensitivity training for juvenile hall staff.

Massachusetts Coalition for Occupational Safety and Health. Working in a mixed-gender program, MassCosh and its Teens Lead @ Work campaign engage low-income teens of color in social justice activity related to their rights on the job. This includes educating them about the social, economic, and political obstacles facing immigrant female teen workers, and helping them work collectively to create strategies for seeking systemic social change. Working to create equal opportunities for both young women and young men, participants raise awareness of the exploitation of young workers by serving as peer leaders, trainers, and organizers. To increase awareness of gender issues in the workplace, Teens Lead @ Work participants engage in discussions about power, the oppression of women, and sexual harassment of women and men.

Increasing financial resources is just one component of strengthening organizations. Working together, CFYS partners will identify other capacity-building areas such as organization and staff development, documentation, staff and youth transitions, and membership development. Through annual partner gatherings, conference calls, and targeted one-on-one assistance, CFYS partners will work together to develop these capacities.

Annual Convening

A key venue for progress on the learning agenda is the annual convening. These meetings are opportunities for all partners to come together, step outside of their day-to-day work, share strategies for social change, and learn from each other's experience. Key to this is the conscious effort to build partnership between and among donors, youth organization staff, and youth. The first annual convening was held in June 2003 in Colorado Springs, Colorado. This convening gathered Ms. Foundation staff, 11 donor partners, 24 program partners, 24 youth partners, and guests for two days of plenaries, workshops, team building, and fun activities. Titled, "Changing the World with Youth in the Lead," the convening modeled CFYS principles, including shared learning, value in diversity, and recognition of peer expertise among the donors, youth organization staff, and youth. Youth and adult partners worked individually and together to design an event that met individual and collective needs. A first order of business was creating an environment where all were welcome, developing a common understanding and language for the work, and crafting opportunities for participants to get to know each other and the work of each organization. Plenaries and workshops focused on programmatic issues such as popular education, community organizing, and leadership development. In addition, sessions on organizational development explored new models of intergenerational power sharing in fundraising, evaluation, board development, and management issues. The focus on intergenerational work, especially modeling what we mean by youth-led, was apparent in the issues and approaches discussed at the convening as well as in how the convening was designed. Three examples include:

Youth challenge and create approaches. Youth leaders, youth organization staff, and donor partners had opportunities to talk about how to create a shared assessment of the issues facing their organizations, communities, schools, and workplaces. The partners also strategized about ways that youth lead social change work and how approaches vary by community.

Youth teach and learn. Youth had multiple opportunities through a youth advisory planning group to shape the agenda and design the meeting. Youth partners led presentations, workshops, and team building exercises, individually, and in conjunction with program staff.

Youth partner with adults. Youth, youth organization staff, and donors had a unique opportunity to share information and strategies with peers and other CFYS partners. The great diversity of participants—varied race, ethnicity, gender identity, age, geography, CFYS role—provided for a rich, learning environment that used everyone's expertise.

The convening presented opportunities to talk about the theoretical underpinnings of the work as well as engage in the pragmatic exchange of a variety of models, experiences, and lessons learned. Most importantly, the convening modeled how CFYS hopes to build a true partnership among youth, donor partners, and staff of grantee organizations that strengthens and enhances youth-led social change work.

Intergenerational Leadership

To ensure that youth are successful in leadership positions, they need to be exposed to varied approaches, receive support for their unique leadership style, and have opportunities to learn from experience. In thriving youth development and youth-led social change organizations, adult staff balance teaching and mentoring with providing youth with training and opportunities to make significant decisions. Teaching takes place between youth and other youth, and between youth and adults. For example, to fully embrace leadership, youth benefit from inclusion in the decision making positions within organizations, through service on boards, finance and budgeting committees, and planning efforts. Our understanding of successful intergenerational leadership will be refined as the initiative unfolds.

To assist groups in assessing their approaches to leadership development and the roles that adults and youth play, CFYS partners developed a series of questions used on site visits. Some questions included:

- How are youth involved in decision making in the organization and program?
- What are some challenges to youth leadership?
- How many youth are in your core leadership team? Who tends to be in this core leadership team? Is there rotation? What are the similarities and differences in the participation of girls and boys, racially diverse youth, and older and younger youth?
- How are youth involved in the organization (board, advisory group, part-time staff, paid interns)?
- What is the balance of adult supervision and youth autonomy?

Participants in CFYS address intergenerational leadership in a variety of ways, including:

Asian Immigrant Women Advocates. Working with Asian youth in Oakland, California, this organization's Youth Build Immigrant Power Project (YBIPP) uses multiple strategies to cultivate leaders. Peer-to-peer leadership development occurs through youth-led trainings and workshops. Adult-to-youth leadership development occurs through adult staff and community members helping youth identify issues of concern to the community and crafting action campaigns. An example of intergenerational leadership occurred as youth identified the issue of the poor working conditions for immigrant women workers as a key community issue. Youth took leadership in raising awareness of the issue among community members, and worked with immigrant women workers to make demands to improve workplace conditions.

Sista II Sista. This organization's Sistas Squads foster young women's growth in public speaking, facilitation, and mentoring. This increased capacity lends itself to peer-mentoring and leadership development as program participants support other young women new to the program. Sista II Sista uses a collective leadership model where girls, young women, and adult women are equally involved in planning and decision making for the organization. Currently organizing to fight violence against women of color, young women in the program have raised awareness of the issue and challenged the police about their sexual harassment of young women of color in Bushwick, New York. These young women provided opportunities for other women—young and old—to talk about sexual harassment, their desire to see the violence stop, and how they can be involved in creating a solution. They have created intergenerational work teams to provide alternatives for women to turn to in cases of interpersonal violence.

Young Women's Project. Based on the experiences and input of young women leaders, the Young Women's Project's Teen-Led Projects combine a strong leadership development approach with social justice campaign work. To achieve both individual growth as well as collective action campaigns, young women work closely with peer mentors and adult staff to identify issues, create accountability within the group, and develop campaigns that benefit from the experience and guidance of youth and adults. For example, to improve the lives, rights, and opportunities of the more than 400 youth living in 30 Washington, DC, foster care group homes, teens and adults worked together to identify the problem, develop draft regulations, and conduct outreach to alert teens in group homes to their rights. Adults provided guidance to youth on how to craft the campaign, influence decision makers, and develop information and training curricula. Youth provided leadership to youth in foster care by developing information and training to help them know and act on their rights.

Colorado Progressive Coalition. Launched by two high school students who appreciated the mission of the Colorado Progressive Coalition (CPC) and wanted a greater role for youth, Students 4 Justice (S4J) works to raise awareness of issues affecting youth and communities of color, trains youth on campaign strategy and analysis, and provides opportunities for youth to engage in social change. To support the key role of youth in the organization, youth, with the support of CPC staff, developed an antiageism training for adult staff and board members. In addition, youth representatives from three high schools serve on the Board of Education, and youth take the lead in developing and implementing awareness and action campaigns. Youth surveyed 350 of Denver's Eastside residents and released a community report on racial profiling that resulted in the passage of one of the nation's toughest antirace discrimination laws. They are also working to remedy racial tracking of students of color that dissuades them from college track classes and investigating the allocation of funds between public education and the construction of a jail costing over $100 million.

Learning Component

The CFYS learning component will engage all partners in a process of identifying, discussing, and documenting promising practices among organizations engaged in youth-led social change work. To this end, CFYS partners have agreed to participate in a learning component, which will include the documentation of capacity-building and learning activities.

While CFYS partners began to ask a number of questions in the early design stages and at site visits to select our grantee partners, the joint learning agenda is established with the active input of all partners, including youth organization staff, youth participants, and donors. The first national convening provided the opportunity to identify themes that will be narrowed down to questions over the next few months. Key questions include the following: What are effective models for shared adult-youth leadership? How does youth-led social change look different in diverse communities and locations? And, how do gender and other identities impact youth-led social change work? Future convenings and site visits will provide opportunities to deepen shared learning on selected issues and questions.

Conclusion

The Collaborative Fund for Youth-Led Social Change provides an exceptional opportunity to explore new directions in the fields of youth development and youth organizing. By developing a unique partnership among youth, youth organization staff, and donors, we can explore program models that take youth identities into consideration and build the capacity of youth-serving organizations. CFYS will document lessons learned and share these findings with the field. Most importantly, by bolstering the practices of youth-serving organizations, youth leaders from across the country will be recognized and respected for their contributions to their communities and the youth fields.

Note

This chapter was originally published in 2003 as "Power and Possibilities: Collaborative Fund for Youth-Led Social Change" by the Ms. Foundation.

References

Cahill, M. (1997). "Youth Development and Community Development: Promises and Challenges of Convergence." Paper presented at a meeting sponsored by The Ford Foundation and International Youth Foundation

entitled "Community and Youth Development: Complementary or Competing Priorities for Community Development Organizations."
Camino, L. (1995). Understanding intolerance and multiculturalism: A challenge for practitioners, but also for researchers. *Journal of Adolescent Research* 10 (1).
Innovation Center for Community and Youth Development. (2000). *Youth leadership for development: Broadening the parameters of youth development and strengthening social activism.*
Lewis-Charp, H., Soukamneuth, S., and Yu, H. C. (2003). "Filling the Gap: Intersection of Civic Activism and Identity in Youth Development." Presentation of Research Findings for the Innovation Center for Community and Youth Development.
Mead, M. (2001). *Gender matters: Funding effective programs for women and girls.* A publication for Women and Philanthropy. Report. Medford, MA: Tufts University.
Ms. Foundation for Women. (2002). *The New Girls' Movement: Charting the path.* The Collaborative Fund for Healthy Girls/Healthy Women Project. New York: Ms. Foundation for Women.
Pittman, K. (1991). *Bridging the gap: A rationale for enhancing the role of community organizations promoting youth development.* Report. Center for Youth Development & Policy Research, Academy for Educational Development.
Roach, C., Yu, H. C., and Lewis-Carp, H. (2001). Race, poverty and youth development. *Poverty & Race* 10 (4).
Zeldin, S., Kusgen McDaniel, A., Topitzes, D., and Clavert, M. (2000). "Youth in Decision-Making: A Study on the Impacts of Youth and Adults in Organizations." Report commissioned by The Innovation Center for Community and Youth Development, National 4-H.

Chapter Nine

Young Women for Change

The Michigan Women's Foundation

Young Women for Change® (YWFC), a girls-as-grantmakers program, is a committee of high school young women, ages 14–17, of diverse backgrounds who assess the needs of girls and young women in their community and grant funds to nonprofit organizations working to serve those needs.

Studies show that young women continue to face gender-specific challenges which impact their economic self-sufficiency. Research also confirms that a majority of girls express and develop their own opinions freely and take advantage of leadership opportunities more fully in gender-specific situations. However, gender-specific economic and leadership opportunities are few and far between. Less than 7 percent of philanthropic dollars go to programs which specifically address the needs of girls and young women.

History

Michigan Women's Foundation (MWF) is particularly proud of the YWFC program that was created to have a positive impact upon and improve the futures of girls and young women in Michigan by developing a model for grantmaking which could be replicated in other communities that actively involves young women in philanthropy.

YWFC began in 1996 in Kent County through a partial endowment and expanded to Metro Detroit in 1998. Over the next five years the program expanded to four other sites in Michigan. Every year each site awards $20,000 in grants, ranging from $1,000 to $5,000, to programs that primarily serve disadvantaged girls and young women in their respective communities.

Program Goals

The overarching program goals are:

- To promote systemic change that expands the girl's potential and capacity of for leadership.
- To prepare a new generation of young women for the reins of leadership in philanthropic endeavors.
- To increase the pool of money designed especially for girls.

Program Objectives

To Actively Involve Young Women in Philanthropy

- To help young women reach an understanding of and commitment to philanthropy.
- To create an outlet through which young women may express their opinions, make use of their talents and experiences, and have a positive effect on the lives of girls and young women in their community.
- To make the community aware of the Young Women for Change (YWFC) committee and young women's potential in making significant contributions to their community.

To Meet the Needs and Improve the Lives of Girls and Young Women across the State

- To grant money to the programs and organizations effectively meeting the varied needs of this population.
- To pave the way for YWFC to improve the lives of girls and young women in their own additional, creative ways (for instance, implementing their own programs, voluntarism, et cetera).

To Create a Model for Grantmaking Which Can Be Replicated in Other Communities

- To accomplish two years of effective grantmaking so that success and challenges can be assessed and addressed for future years.
- To establish a curriculum for the initiative to be followed by those implementing similar programs in their respective communities.

To Promote Skill and Leadership Building

- To establish a committee of young women that works together effectively and efficiently.
- To develop the leadership skills of YWFC members.
- To teach and utilize critical thinking skills.

To Promote Knowledge Building

- While incorporating their own knowledge, to educate YWFC members in a variety of girls' issues.
- To expose YWFC members to a variety of programs and people working to serve girls and young women across the state of Michigan.
- To help the Michigan Women's Foundation learn how to more effectively meet the needs of girls and young women through grantmaking.
- To provide a forum for idea exchange and resource building in girls' needs and issues, including developing funding priorities and identifying effective girls' programming.

How It Works

YWFC committees meet for three hours once a month from September to June. In addition, committee members go on site visits to potential grantees and are involved in other leadership and community service activities. YWFC members execute every aspect of the program, from selecting new members and creating funding guidelines, to managing the grantmaking process.

Young women learn to be leaders and decision makers through awarding $20,000 to local nonprofit organizations with programs that serve girls and young women in their community. Priority is given that address the YWFC funding priorities which are set annually. The request for proposals (RFP) is released in November and due in February. Funding decisions are made in May and recommendations are presented by the young women to the MWF Board of Trustees in June. Historically, YWFC funding priorities include social issues which affect girls/young women such as gender equity, cultural awareness, teen sexuality, substance abuse treatment and prevention, peer pressure, dating violence, leadership and self-esteem building, and economic well-being.

Curriculum

YWFC actively involves young women in philanthropy while educating them about the social, psychological, and economic forces that affect their lives. Young women learn collaboration skills as they help the Michigan Women's Foundation address the issues facing girls and young women in Michigan.

The curriculum structure is the same for all Young Women for Change sites, with the materials for the different pieces of the curriculum being tailored to each group as necessary. The main components of the curriculum are:

- *Icebreakers.* To create a strong group dynamic of respect and support; laying the foundation for team work.
- *Leadership.* To develop the girls' individual strengths, confidence, and commitments to make a positive difference.
- *Diversity.* To enhance the girls' understanding for diversity in race, ethnicity, gender, and thoughts, both in the group and their surroundings.
- *Women in philanthropy.* To discover past motivations and trends in women's philanthropy and to explore the girls' present and future roles as philanthropists.
- *Issues affecting young women.* To examine of the needs of girls/young women in their community.

Grantmaking Process

- Set Funding Priorities
- Review Proposals
- Conduct Site Visits
- Determine Funding Recommendations

Benefits for Participants and their Communities

Not only do the young women gain a valuable learning experience from the program but also there are hundreds, if not thousands, of other girls and young women throughout Michigan affected by the work and dedication of YWFC participants. Hands-on learning experiences that reach far beyond the four walls of a classroom are essential to the growth and development of young people. Nontraditional learning opportunities for young women to become actively involved in and learn about philanthropy at an early age are limited or often nonexistent.

The Michigan Women's Foundation is committed to fostering economic power leading to a philanthropy philosophy of long-term community involvement and leadership among young women. This is accomplished in partnership with organizations businesses, individual donors, and other foundations in order to create additional spaces for young women to become active participants in the grantmaking process.

Building positive relationships, working with diverse individuals, utilizing problem-solving skills, adapting interpersonal skills to accommodate a variety of ages, and communication skills are all examples of lifelong tools YWFC develops. By learning about different organizations dedicated to increasing opportunities for women and girls, YWFC helps participants deal with the numerous barriers women and girls face on a daily basis. Most importantly, they facilitate social change.

Young Women for Change Alumni

YWFC actively involves young women in philanthropy while educating them about the social, psychological, and economic forces that affect their lives. By being grantmakers, young women review organizational budgets, learn to ask critical questions, practice consensus decision making, and gain valuable life experience. Through this opportunity young women become active members of the community, leaders of tomorrow, and philanthropists today.

Six past YWFC participants share how the program opened their eyes to their community, to other young women with diverse ideas and the reality that they could make a difference:

Sarah Howie of Grand Rapids, MI (1998–2002). Sarah is currently a student at Colgate University in Hamilton, New York where she is majoring in international relations. She spent one semester living in Russia with one of the country's most noted female poets and worked for a public relations firm. "Young Women for Change was an amazing opportunity for me," Howie said. "YWFC was a way for students to learn about the problems that exist in our community." "The YWFC experience gave me the expertise to get involved at Colgate in the University's Budget and Allocations Committee that awards funds to student groups for campus programs. Allocating $20,000 was good training since now I give out more than $500,000 each year for student lectures, health fairs, concerts and other activities."

Melody Moore of Detroit, MI (1998–2002). Melody is currently a writer for a newspaper in Detroit, runs her own communication firm and attends Wayne State University, where she is majoring in journalism. Moore states, "YWFC is a good learning experience. It provides one with a level of professionalism, knowledge about grant writing, philanthropy as an industry and it will broaden one's awareness about true diversity." "I learned how to give wisely," Moore said. "I learned how to be a team player. As the committee moved through the allocation process we went through some tough times in terms of making decisions about which programs should be funded. We were opinionated. We worked well together, in spite of our differences."

Margaret Sweeney of Grosse Pointe, MI (1998–2000). Sweeney graduated from the U.S. Naval Academy in 2004 with a degree in oceanography and is currently working on a master's degree. She was commissioned an Ensign in the U.S. Navy and is beginning her career as a Surface Warfare Officer. "Active participation in YWFC helped me learn more about myself and my ability to work with others from different social and economic backgrounds," Sweeney said. "I became keenly aware of not only the problems facing young women, but the organizations that are making it their mission to find solutions, inspire and empower those who need assistance."

Emily A. Malloy of Grand Rapids, MI (2000–2002). I became involved in Young Women for Change because I was looking for a way to become more locally engaged with women's issues. When I heard about YWFC, I was thrilled for a number of reasons. YWFC helped me not only to learn about issues affecting women and girls in my community, but it also empowered me to impact their lives in a tangible way through grantmaking. At the time I interviewed for YWFC, I was (unsurprisingly) unsure about my purpose in life . . . I was a junior in high school and losing a lot of sleep about how I could make the most impact with my little life. I was hoping that YWFC would be able to provide resources, or at least a sounding board, for some of my career questions. My first "resource" came from something unexpected. The Grand Rapids offices of YWFC are housed in United Way Center. When I entered the building for my interview, I could only stand in the doorway and look up at the thoughtfully restored downtown building and think, "I wouldn't mind working *here* someday." After two years in the YWFC program, I left for college, but not without carrying the empowering lessons of YWFC with me. I went on to be the editor-in-chief of the largest yearbook in the country, was appointed to the Michigan Women's Commission by Governor Granholm and continued to be involved in other community roles . . . all leadership roles that I was undoubtedly prepared for because of YWFC. And on a Monday morning last August, three months after college graduation, I began my first full-time job. And I was standing in the doorway of United Way Center, in downtown Grand Rapids—about to become a marketing associate for Heart of West Michigan United Way. It seemed to be such a twisted journey to get right back to where I had started from. But it all made sense. I was once the girl who was going to join the Peace Corps. I thought going to another country was the only way to *really* make a difference. But YWFC made me excited about the potential for change right in my own backyard . . . I've never thought twice about my decision to return to my hometown and work for change in the community where I had my start. My husband and I even bought a house right in Grand Rapids—just to make it all a done deal, I guess. It was truly the last thing we expected to do just months after college, but it speaks to why programs like YWFC are so important to our communities. When you empower young people like YWFC does, you do more than educate them about philanthropy in their community. You invite them to a lifelong dialogue about that community. Whether you realize it or not, you encourage them to come back and buy a home in a neighborhood in that community. Because of YWFC, my passionate concern for West Michigan will be long-lasting. (I have a 30-year mortgage to prove it.) And the best part is that my story is an incredibly simple one compared to the others from YWFC program participants.

Tessa Haviland of Kalamazoo, MI (2002–2005): Being a part of Young Women for Change gave me confidence that young women can make a difference in their world. Every year we received more applications for grants than we had money to give out. This showed that it was more than just young women who saw a need for change in our community. I have not met anyone like the young women I worked with in Young Women for Change. They understood and saw inequality and instead of complaining, they worked to change their world into something better. Since I have been at college (Kalamazoo College) I have met others who wish to change the world, but no one who has had the resources to make a change. Even more than changing the world Young Women for Change brought me to a new understanding of feminism. Everyone within our group was different and each young woman understood feminism and understood issues facing young women in a different way. This demonstrated, for me, the incredible difference among young women. There are so many issues in our world and everyone understands them differently, but that does not make any single opinion less valid. The idea of cooperation, of allowing all opinions to be heard without giving one idea more importance, has allowed me to work with other groups (both men and women) to make a difference in their communities. I will take Young Women for Change with me wherever I go. I will never forget the experiences, but more importantly the women, who showed me how to change my world.

Bethany Neigbauer of Ann Arbor, MI, (2003–2004): I was a member and the secretary. During my year in the program I learned so much about how much of an impact I could really have on the community. I never imagined I could help so many people. The Ann Arbor community is my home and I love to give back anything of what the community has given to me. Young women for change made me believe in myself and see how much power a young women in society could really have. It's an experience that I will never forget and hopeful help me in my future community service activities. With regards to where I am now, I will be graduating from Pioneer High School in June and will be going for my bachelors degree in Architecture at the University of Michigan. I have already been accepted to the school of Architecture for my junior year and can't wait to go. I will be living at the dorms and possibly trying out for the golf team. I played Varsity golf for Pioneer for three years, and haven't made my decision about playing collegiate golf yet.

The Michigan Women's Foundation

Vision

Women empowering women and girls to realize their dreams of economic self-sufficiency and social equality resulting in a better society for all.

Mission

MWF is the only statewide foundation focused solely on women and girls. MWF develops emerging women leaders and provides financial and technical assistance to nonprofits, creating skills and leadership opportunities.

The Michigan Women's Foundation believes in

... strong women
... being change agents
... giving back
... empowering others
... equity

... the courage to take risks
... being role models
... understanding and acknowledging our power
... serving women and girls
... personal responsibility

We believe that women and girls continue to face significant barriers and challenges to reaching their full potential. Therefore, the Michigan Women's Foundation promotes the economic self-sufficiency and personal well-being of women and girls in Michigan. We do this by:

- providing assistance and funds to nonprofit organizations serving women and girls;
- by educating the general public, policymakers, and donors; and
- by encouraging women and girls to exercise their responsibilities as philanthropists.

Core Values

In following the mission statement for the Michigan Women's Foundation, seven core values were established, around which MWF operates:

Integrity. Committed to honesty and fairness in all of our activities and respect for all of our stakeholders.
Empowerment. Inspiring independence, thus enabling women and girls to achieve their full potential.
Equity. Assuring justice and fairness to those that we serve and for all who work for and with us.
Leadership. Demonstrating responsible risk taking, as well as supporting innovative initiatives and programs.
Sisterhood. Developed through our common identity and bonds.
Compassion. Acting with respect and empathetic understanding to improve the lives of women and girls in Michigan.
Quality. Expecting that all we do and all we support will adhere to the highest standards of quality.

Note

This chapter is based on a Publishing entitled "Young Woman for Change" by the Michigan Women's Foundation.

Chapter Ten

Collaborations for Gender Equity in the Context of Policy and System-Wide Change: An Interview with the Editors

Shirley Mark, Alice Ginsberg, and Marybeth Gasman

The following chapter is presented as an interview between the editors of this book and Shirley Mark, former program manager at the Schott Foundation. As noted in the introduction to the book, Schott's work is singled out here because in their early initiative Gender Healthy/Respectful Schools, they brought grantees together on a regular basis to share their ideas and experiences and pose questions. Although Mark is no longer at the Schott Foundation, the Foundation continues to keep gender as a major category of funding, including a unique and special initiative (described in section one) on *Saving Black Boys*.

The Interview

Editors: What was the impetus for The Schott Foundation to design an initiative focused on gender equity in education?

Shirley Mark: The founders of the Schott Foundation were very concerned about the research that revealed gender bias in education and its effects on girls' development, academically and with regard to their identity and self-esteem. At that time there was extensive media coverage on boys' isolation and on bullying and teasing. We later came to a deeper understanding of gender bias and realized that gender bias impacts boys as much as girls, in different ways. The Schott Foundation wanted to impact systems of oppression. The funding was prioritized for public schools because that is where children spend the bulk of their time during childhood and where gender bias can be most detrimental to the long-term education and overall well-being of girls and boys.

Editors: What issues/questions did you hope to address?

Shirley Mark: We wanted to change the way educators and school systems treated children/youth based on their gender. We are all socialized with ideas about gender, about what girls should or shouldn't do, what boys are good at or not, et cetera. "Boys will be boys" is a common phrase that condones socially undesirable behaviors. "Girls aren't good at math or science" is a socially-accepted stereotypic assumption about girls. The Schott Foundation had a vision for changing the way teachers teach so that all children would be encouraged to develop and pursue a wide range of academics, sports, the arts, and other subjects, to their fullest potential, without regard or preference due to their gender. We wanted public school environments to demonstrate inclusive, positive, and respectful attitudes so that children from a variety of backgrounds would feel supported. Urban public schools, often with high numbers of students of color and high poverty rates, were a priority. We were concerned about race, socioeconomic class, sexual orientation, and other oppressions that affect the development of young people and we expected the grantees to address these issues in their work.

Editors: Who did you involve as you designed the initiative (for example, Schott staff and board members, a gender-focused advisory board, potential grantee organizations, gender experts, teachers, administrators, students, parents et cetera)?

Shirley Mark: As the Gender Healthy/Respectful Schools program was being developed, we consulted with numerous local and nationally renowned educators and experts. We held focus groups in the Boston and Cambridge public schools as well as with experts. Some of these experts included Peggy McIntosh, Beverly Daniel Tatum, David Sadker, Carol Gilligan, and many others. We also sought guidance from women's organizations and philanthropic networks. Janie Victoria Ward conducted a literature review exploring questions such as:

- What are the educational gaps based on gender?
- What are the best practices with regard to professional development?
- What legislation and policies exist (state and federal) that address gender issues, and what is the status of their implementation?

About the same time, I read about a program in the Philadelphia Public Schools, Gender Awareness through Education (GATE), sponsored by the Pennsylvania Humanities Council and contacted the lead staffperson on that project, Alice Ginsberg. She was incredibly encouraging and supportive of the Schott's interests and assisted by organizing focus groups with teachers and consultants involved with the GATE program. Those focus groups essentially shaped what later became Gender Healthy/Respectful Schools.

Editors: What did the initiative look like in its early phases, and how (if at all) did it evolve over time?

Shirley Mark: During each of the three years of Gender Healthy/Respectful Schools, a "Request for Proposals" was issued to every public school teacher in the Boston and Cambridge Public Schools, reaching approximately 140 public schools and thousands of teachers. Through a competitive grants review process, teams of teachers were selected and awarded a grant to support their projects. During the first year, in the interest of stimulating new and innovative work, we decided to take some risks and funded individual teachers without requiring administrative support. We learned much during that first year and decided in the second year to raise the stakes and did require administrative support as well as an expressed interest in changing part of their school culture and/or educational practices. The proposals for the second and third years were more visionary and funded projects proved more effective at creating broader change. With regard to communications, we also learned in the first year that the public did not understand the term, "gender equity." Most people thought it mean girls-only issues. Also, it did not address social behaviors such as bullying and teasing which are gendered issues. In the second year, we changed the name to "Gender Healthy/Respectful Schools."

Editors: What were your primary criteria for grantmaking? How and why was this criteria agreed upon?

Shirley Mark: As I mentioned earlier, the board was very committed to addressing gender bias and its detrimental effects on the development of children. Our vision for the Gender Healthy/Respectful Schools initiative was to support efforts to challenge gender biases so that schools would become positive and nurturing environments for all children.

Editors: One of the things that makes this initiative truly unique is that Schott brought the grantees together on a regular basis so that they could learn from each other and so that Schott could be more genuinely involved. Can you describe this process? How it worked and what it accomplished?

Shirley Mark: We knew that this work was difficult—few educators had challenged gender bias in the schools. Research had generally focused on youth co-curricular programs but seldom in the public schools addressing school culture. In order for these teacher-teams and schools to be successful, they needed professional development, training, individual consultation and day-to-day support. A Request for Proposals was issued for a consulting team to provide this support and the Gender and Diversities Institute of the Education Development Corporation was retained for this service.

During each year of the initiative, the Education Development Corporation convened the 100+ grantee-teachers multiple times during the year. They provided technical assistance and training, individual consultation, and evaluation support. Grantee teams expressed their appreciation of these convenings that provided the opportunity for teams to work together outside of their school setting and also work with educators from other schools to share their challenges and successes with each other.

We sponsored at least one annual conference where national experts on gender and race worked with them. Peggy McIntosh, David Sadker, and Janie Victoria Ward were several of the speakers who spoke at these conferences. In addition, we held an annual end-of-year celebration to showcase their work to a broader community.

> *Editors:* Did you use any special process for evaluating the effectiveness of the programs, or deciding which groups (if any) should get multiyear funding? If so, please describe.
>
> *Shirley Mark:* The Education Development Center assisted teacher-teams in developing an evaluation model for each of their respective programs. The evaluation team led both large-group trainings as well as worked with each of the 20 teams to develop their evaluation model. The challenging part of this evaluation was that it was not one evaluation, but many. Because the grantee teams were so diverse, the evaluation approach also needed to be diverse. For many of the teams, we looked at leadership and professional development, programmatic outcomes, impact outcomes, and sustainability. For teams addressing academic issues, we looked at quantitative data pertaining to academic performance and narrowing the achievement gap, interest in the new content area, willingness to challenge prior assumptions, and attitudes based on gender and ability to reach larger numbers of students.

Chapter Eleven

Sisters Empowering Sisters and a Case Study: Girl World

The Girl's Best Friend Foundation

At Sisters Empowering Sisters (SES), young women are the face of grantmaking. We go head-on with our grantees, who are our peers. This eliminates the power structure often created by (adult) authority figures. At SES, we're not just the face of grantmaking: we're also the brains, hands, and heart. We make the choices, the decisions, the acts, and the reasons. We carry the passion of making positive change in our communities. We make a direct effect on our peers that travels outward and upward.

Sisters Empowering Sisters was created in 1997 as a vehicle to directly incorporate and elevate the voices of young women into the work and vision of Girl's Best Friend Foundation (GBF). GBF was begun in 1994 to help challenge the status quo and support powerful activist girls and young women. GBF has challenged the way traditional grantmaking occurs in many ways—perhaps the most direct way that they have done this is by shifting power, particularly decision-making power. Young women and youth being the sole or lead decision makers in terms of funding is arguably an act of social change in itself. SES involves thinking critically and raising awareness of issues that directly reflect young women's lives and experiences. Grantmaking as a process doesn't often foster that kind of connectedness, and as such, we are doing things differently, freshly. That is the vision of SES—to do things differently. Flipping the script on power and privilege and choices to help young women find their own power and make their own choices.

Current Structure and Philosophies

In trying to understand what the real philosophies behind SES are, *feminist, social change philanthropy* needs to be clarified. We think of philanthropy as the giving of time, talent, or treasure. For bright young women who are being introduced to philanthropy for the first time, this definition proves most effective as far as providing an understanding of the basis of our work: even for our peers who weren't involved in philanthropy before and needed to achieve a more concrete understanding of what embodied.

When we talk about promoting girl leadership, we ask that groups are led by young women/girls. Feminism has unfortunately been tagged with certain generalizations that add to a somewhat mottled reputation. What it's really about, though, is striving for equality among genders—not domination by one or the other. Because SES recognizes the difference between girl leadership and girl dictatorship, we fund programs that buttress the concept of girl leadership, with the understanding that equality is a struggle and goal for all genders. As seen in Teen Talk, a grantee from 2004: it is a high school-based program focusing on sex, sexuality, and what it all means that caters to students of all genders. Teen Talk is led by young women but the active audience consists of different genders, which better suits their subject matter and their goals. This funding concept is a bit new for SES, but one that we believe is really central to the idea of feminist grantmaking. Real change has to include everyone.

Given that SES gives grants to programs by and for girls and young women in order to promote leadership and power among young women, we call that feminist, social change philanthropy because these grants are responsible for creating positive changes within communities of young people. By giving groups and programs that are girl-led and girl-directed the funds that they need in order to ensure that their project can exist, young people in the community are impacted and there is definitely a change in society.

SES believes that grantmaking should be a partnership, not a power trip. We want our current and potential grantees to know that we're all out to make change. The only real difference is that we have funds to give out so that making change can be a little more possible. For this reason, the Sisters incorporate time to meet with prospective grantees into the grantmaking process. At these meetings, called site visits, we get to have a personal conversation, a face-to-face encounter with the girls/young women behind the ideas, and we are able to get more information about their projects, unable to be conveyed just through paper. In other words, we like going past the paper and print to see their passion for and dedication to their project in person.

Besides connecting with grantees, the Sisters are sometimes involved in meeting with other youth groups to share experiences with them and tell them more about our program. Through these gatherings and experiences, many more people become aware of the needs and power of young women, and want to get involved in different ways.

Sisters largely inform and influence the direction and execution of the program, to varying degrees throughout the year. Often, training sessions and other meetings are facilitated by the young women and activities are shaped by their interests. Each year consists of a kickoff retreat and ends with an event where the Sisters are honored and the work they've done over the year is celebrated. SES comprises of a diverse group of young women in both identity and experience: geographically, culturally, economically, and experientially.

Accomplishments and Activities

SES has funded nearly seventy different youth projects since its inception in 1997, including a diverse array of initiatives. Some of the most recent projects we funded are a regional young women's health conference, documentaries made by and for girls looking at questions of nontraditional career opportunities, a group of young queer women compiling a literary 'zine that reflected their own lives and experiences. SES also has a strong history of being directly involved in broader community activism and advocacy. We sent five Sisters to attend the March for Choice (on reproductive rights and justice) in April of 2004.

Sisters Empowering Sisters implies that we help others realize the strength of their voice and the power of their actions. There are many changes that Sisters have gone through as a result of their experiences in SES. Many of the Sisters have grown to be more comfortable in their own skin and are quick to show our friends how liberating it is to live on our own standards and not according to what society refers to as "the norm." The irony in the name of Sisters Empowering Sisters is that while we're all working to empower girls and young women in our communities, we are empowering ourselves and each other in the process.

Because we do grantmaking in the fall, we spend each spring doing a leadership and empowerment project. In 2004, SES conducted a research project analyzing the impact of commercial music videos on the lives of young women. Just this past spring, SES produced a 30-minute documentary called *Respect Me, Don't Media Me!* that both portrays our findings from the research as well as what we as young women can do to change it. *Respect Me, Don't Media Me!* focuses on five specific areas in the media that we think young women are particularly influenced: body image, sex/sexuality, role models/icons, relationships, and personality/style. Just because we're not a media empowerment program, we see no reason why young women can't create our own positive media and social change at the same time!

Impact and Future Plans

Youth—and young women's—philanthropy is important to social change because it makes movements more accessible to young and old alike. It provides a new reality allowing each generation to control its own fate and to have a say in the future of our world. SES believes that the youth of today are not only the leaders of tomorrow; they are the leaders of now.

SES has helped changed the face of grantmaking and youth programming. It not only involves young women, but is also led by young women in order to improve the lives of our peers. From a youth leader in a recent grantee project, "SES made me realize how women face so many different problems in the world and that there are so many ways that you can be involved in helping solve those problems. It's awesome to see all different types of girls and groups and what we're all trying to accomplish!"

Everything within SES is geared toward and made accessible to young women: from seemingly small things like the wording of a request for applications to the fact that we can always get reimbursed for traveling to/from meetings. It is through this deliberateness that we hope to really impact the way youth programming plays out. We just began our eighth year, and are going strong. Girl's Best Friend Foundation, our host organization, is slated to close its doors in 2008—but SES will continue! We are currently in the midst of finding a new, local home that will maintain SES' true identity and commitment to young women's empowerment.

Wherever we end up, we know this: SES flies in the face of the "apathetic youth" stereotype—a passion to help fuels a passion to change. And with young women at the helm, we are flying fast.

A Case Study Program:
Girl World of Alternatives Inc. Project:
Girl Driven Research Project

Jessica Palmert

When I first attempted a participatory evaluation research project with a team of young women I had no idea what I was doing, where the research was going, and what we were going to do with the findings if we ever managed to collect them. Conducting participatory evaluation research proved to be a learning experience not only for the young women, but for our staff, agency, and community. I learned about participatory evaluation research through the trainings facilitated by P. Catlin Fullwood at Girl's Best Friend Foundation. Programmatically I was at a place where I was looking for a project-based learning experience that could challenge the young women I worked with. Many of them already had had experience doing grantmaking, fundraising, planning agendas, and facilitating discussions. I needed a project that could challenge them to develop a new skill set and that could incorporate some social justice topics into the curriculum. A participatory research project seemed like an excellent way for the young women to identify and explore the issues that were most important within their community, continue to build their leadership skills, strengthen their critical thinking skills, offer an opportunity to work on employment skills, and indicate if our programs were addressing the most crucial issues for young women today.

Research as Opportunity

In order to engage the young women already involved in Girl World to take on another responsibility and a six-month commitment I presented the researcher positions as paid employment opportunities. A chance at paid employment is what initially made the project attractive to our program participants. The second year the applicants had seen what the first year team accomplished and were interested in being "experts" on a topic interesting to youth. Due to the success of the first project the second year team felt the role of research was prestigious and wanted the recognition that came with presenting their findings to the community. The second year and third year participants came to the project with ideas they were interested in exploring and expected the project to be challenging, but fun.

Generating the Topics

Our research topics have always been generated by the young women. The only perimeters that the staff set were that the topic needed to be about youth. The first year we asked the team to only interview girls but the second and third year teams wanted the male perspective on their subjects and males were therefore included in the surveys and interviews. Staff originally wanted the youth to explore a "community issue" in the sense of safety of girls in the neighborhood, girl on girl violence, or sexist billboards in their neighborhoods. The first and second teams were more interested in exploring personal politics and subjects that were concerned with their racial and cultural identities. It wasn't until the third year that the young women arrived at the topic of girl on girl violence, which was a lesson for the staff to learn. The young women will arrive at topics in their own time if they are given the space and freedom to do so.

Naming the Research

The first year the young women chose "The Hair Project: How do African American Young Women Feel about Their Hair and What Influences Them to Feel That Way." This subject allowed the young women to examine the historical, political, and personal significance of their hair. The team asked their peers whose hair they idealized, what they wanted to change about their hair, how much they spent on their hair, what hairstyles are considered "professional," and who influenced their hairstyles. The second year the team was interested in exploring "Why Do Teens Desire High Priced Material Items." This project probed into the importance teens place on owning brand name items.

The third year focused on girl on girl violence and the contributing factors to the conflicts. Some of the main findings revealed that boys, power over other girls, and competition regarding looks are all factors that contribute to girl's conflicts with one another. The media also plays a role in girl on girl violence by promoting it through talk shows, movies, and ads and portraying female conflicts as trivial, sexual, and funny.

Getting Started

After securing funding from GBF and technical assistance (in the form of a graduate student who helped us to navigate the research process that first year) from Loyola University's Center for Urban Research and Learning we were ready to create our research team. Staff handed out job descriptions and applications to interested young women and each applicant had to interview with staff and previous team members. After the team was selected they underwent about six weeks of training by both staff and members from the previous research team. The youth trainers developed curriculums around the aspects of the research they found to be most challenging the year before. The youth trainer's insights and coaching have proven to be very valuable for each new team. During the trainings the young women are introduced to research basics such as creating a hypothesis, data collection methods, confidentiality and ethical research, and qualitative/quantitative coding and analysis. The team also worked on teambuilding and engaged in a number of social justice workshops that investigated the issues of racism, classism, adultism, and sexism.

Making It Real

Every year the training topics are explored through interactive activities that include role plays, debates, media critiques, articles, films, and games. One of the activities that is meant to illustrate classism in society is the "social class dinner." In this activity every team member is given a role in a restaurant from the owner to the dishwasher. The group engages in a role play where everyone functions in their role and helps to prepare, serve, and clean up a meal for two restaurant goers. At the end of the meal everyone is paid for their duties and allowed to "purchase" some of the leftover food with their earnings. The owner can buy the most, the waitress some, and so on. As it turns out the bus girl can only afford a tiny portion even though she worked as hard as everyone else. The exercise is always thought provoking for the young women and inevitably someone becomes very frustrated when she realizes how unfair the allocation system is.

Practice Is Critical

One of the youth trainers created an activity to practice interviewing skills because she had really struggled with them the year before. In the activity one team member was the "rambling interviewee," one was the "confused interviewee" and one was the "quiet interviewee." Each of the interviewees was paired with another team member who had to be creative when dealing with the particular challenges that this interviewee presented. This activity allowed the team to practice their interviewing skills and work out the kinks in their questions. After practicing they realized that some of the questions were not specific enough or difficult to answer and they then changed them before they went out in the field. One of the young women reflected how challenging the interview process was when she said, "The most challenging part in doing my research was the interviews. Most of the people were very vague about their answers, and I had to keep asking them to explain or go more in depth." Making sure that the youth get a chance to practice and build their interview skills is critical to their success in the field.

Deciding the Questions

When we arrived at the point in the research process when we were ready to create the questions we asked the team to brainstorm as many questions as they could about the topic. Each team started with a pretty general topic so the brainstorming process usually served as a time when they narrowed down the subject to what they really wanted to know. The first team started out talking about beauty and was interested in understanding why it was so important to young women. We drew a woman's figure up on the board and started at the top of her head and worked all the way down to her toes talking about all the ways she could "beautify" herself. We came up with over 150 things she could do to change her physical appearance in order to be more beautiful. One of the young women remarked, "It seems as though the natural black woman is used tat (as) the definition of what is ugly and what needs to be changed in order to beautiful." These kinds of "ah ha" moments during the creation of the research question are vital to making the process real for the young women. The conversations and exploration of ideas are essentially to the research success and it is therefore imperative that the staff allow enough time for the concepts to fully develop and evolve. Staff had to take the time to ask the young women what are the most important issues in their community? What are the most important issues to young women? From these types of probing questions the team brainstormed possible research questions and then worked through the consensus model in order to choose their final question.

Deciding on the Methods

Every year during the training sessions we introduced different types of data collection methods to the young women. We discussed how the information would differ if we used a questionnaires versus an interview or observation. Each team member practiced a particular method and shared their findings with the group. When we are sure everyone is familiar with the options the team will choose a method they think will be most effective for their question. After the data collection methods were determined it was necessary to practice the method a few times before the youth were sent out in the field. For photo journaling we shot at least two rolls of film and had them developed before we conducted the formal research project. Critiquing the pictures and talking about ways to capture the information more dramatically or to try another angle really allowed the girls to get confident with the method before the real thing. Interviewing was another skill that we practiced in a number of different ways. We had each member practice doing mock interviews in front of the group and everyone participated in a critique of the interview at the end. This allowed the group to see interviewing techniques that worked and to be involved in a group learning process.

Phases of Data Collection

Data collection proved to be both an empowering and challenging process for the teams. A number of times the young women realized how they should change their data collection methods to make them more effective or realized what questions they should add in order to obtain the information they were after. By the third year in the process we changed the data collection process so that the team collected one piece of their data completely before they decided upon and designed their second collection method. Photo journaling was a trying process for many of the girls as they would bring their film in and look at what they actually captured. Many of the photos were unusable and sometimes the young women would get discouraged. One unexpected complication to the project was that different girls excelled at different data collection methods. One of the girls had a very difficult time approaching her peers and interviewing them, but she took amazing photographs. In an effort to adjust to the varying skill levels we allowed the team members to choose the methods they were most comfortable with and as a result we got back much better photographs and more in-depth interviews.

Keeping the Focus—What's the Point?

During the first year it was very difficult to navigate exactly where the project was going and what we were going to do with the research findings once we obtained them. I remember at one point toward the end of the project one of the young women was completely exasperated when she realized there was not a predetermined use for their findings. "You mean we are doing all of this work for nothin?" she proclaimed. At the conclusion of the project we realized that the research findings help the staff evaluate if our program is addressing the juxtaposition of racial identity, youth culture, self-esteem, and gender politics, which are the issues that the young women are identifying and documenting as the most pressing.

Using the Skills in Everyday Life

In a very concrete way conducting research increased each young woman's critical thinking and leadership skills. As the project has progressed over the years we have come to understand that the process of creating critical questions, interviewing peers, analyzing data and media messages, and presenting research findings are all incredibly powerful and transformative. It is transformative because it helps them to develop and explore critical questions in their lives. One young woman summed this idea up when she said during her final evaluation, "I learned to analyze more things, everything is much deeper now there is always a cause and an effect, I see things in a different way."

Another concrete example of this transformation occurred with a young woman who was frustrated by the administrative staff at her school who she felt was the targeting and handing out unfair disciplinary actions against the students of color over dress code issues. Once she had learned the basics of the research process from our project she decided to design her own data collection method in an effort to capture the targeting of students of color and then use the data to confront the administration. Seeing this young person empowered to ask critical questions in other areas of her life and to have the tools necessary to explore those questions was really a testament to how empowering participatory research can be for young people.

Change Takes Time

Even though we have seen a transformation of thought and deepening of critical thinking skills it is important to understand and acknowledge that participatory research does not necessarily lead to an immediate change in actions. We have seen varying degrees of change over the years but we do not believe that change is the only indicator of a successful research project. After exploring the political and historical significance of African American hair two of the young women on the team decided to stop the chemically straightening their hair. The other two choose to continue straightening but felt that now they were making an informed decision. The second year none of the team members decided not to buy brand name clothing even though they knew it was incredibly expensive, and that people were prioritizing brand names over paying bills and college. In this third year two of the young women stopped fighting other girls, but another member still believes that girl on girl violence is entertaining. One of the young women summarized her change in consciousness when she stated, "I found that I am upset that many girls know that they shouldn't be fighting especially over boys, yet they do and the boy's answers were even worse. I often find myself being a mediator now. Fighting is degrading and can be avoided, there are alternatives to such acts, but people need to realize that."

Change happens at different times and at varying paces for all people and research does not guarantee change, but it does facilitate a consciousness raising process that hopefully plants seeds for future reaping.

Note

This chapter is based on a publication entitled "Sisters Empowering Sisters" by the Girl's Best Friend Foundation.

Chapter Twelve

Gender Equity in Urban Education: New Relationships between Funding and Evaluation

Alice Ginsberg

This chapter focuses on the evaluation of programs and policies designed to promote gender equity in schools and to raise awareness in general about gender issues in urban education. It raises the following problems:

- How to frame and define gender as an important issue in urban education.
- How to decide where and by what criteria to distribute limited resources to gender-based educational programs in urban communities.
- How to *assess* and *evaluate* the impact and importance of gender education work both locally and nationally.

The chapter has two main themes: The first is *how* such programs are typically assessed, meaning the different tools and criteria that are used to make judgments as to their worth and ultimate "success," or "failure." The second theme addresses the problems inherent in *funding* programs which are explicitly focused on gender and gender equity in urban education. Some recent studies and reports (Grady and Aubrun, 2000; Mead 2001; Three Guineas Fund, 2001) suggest that it is still difficult to convince key educational stakeholders that gender equity in education is important—despite over a decade of research showing the impact of gender bias on both girls *and* boys throughout their education (AAUW, 1992; AAUW/Research for Action, 1996; Davis, 2000; Francis, 2000; Ginsberg, Shapiro and Brown, 2004; Leadbeater and Way, 1996; Orenstein, 1994; Sadker and Sadker, 1994; Shapiro, Sewell and DuCette, 1995; Ward et al., 2002). Skepticism as to the worth of gender equity programs is even greater in the case of *urban* education, as many of these students have been labeled "at risk" and schools are encouraged to focus on teaching nothing but "the basics" (for example, learning to read) ignoring differences in experience, culture, language, and/or community interests and values.

It is important to look at the issue of gender equity from the perspective of the teachers and educators actually working in classrooms and schools as well as that of the funders, administrators, and policymakers who are often considered "outsiders" to the reform process. Ultimately, these groups must work together, share common goals and language, and produce evaluations that support innovative and effective programming for urban education. This is not, however, an easy process.

Recent studies confirm that gender equity is not a priority, or even a visible issue, for many school reform advocacy groups, despite an interest in broader issues of equity and racism (National Center for Schools and Communities, 2002). The general public is also extremely confused about just what gender equity in education means, and, in particular, conflicted about whether paying closer attention to girls somehow means *shortchanging* boys (Grady and Aubrun, 2000). Moreover, it has been found that most educational foundations are apt to seek the least controversial funding criteria, that is, to fund programs that are broadly considered to be "universal," rather than targeting one specific group of children (Mead, 2001).

While educational program developers must constantly try to "prove" their programs' worth, many program funders are also looking for ways to justify funding decisions that target scarce resources for school reform work with gender at the center (Mead, 2001). In the current educational climate of high-stakes testing and standardized curriculum, increased accountability, and widespread systemic reform, many innovative reform programs designed to raise awareness about gender issues in schools are short-lived, underfunded, and isolated from other reform initiatives (Ward et al., 2002). This is an especially relevant issue when thinking about evaluation, as program evaluations are (often) fueled by the desire to gain *future* funding, and funding agencies routinely use evaluations to make important decisions about what kinds of programs and which programs they will support.

This paper addresses some of the ongoing questions and concerns that foundations and other educational funding agencies grapple with as they try to support such work. After briefly exploring some of the different models of gender equity programs in section 1, I look broadly at the issues of educational evaluation and of accountability (section 2).

Though the questions raised in this section are not all specific to gender, they do raise important questions about why we place so much emphasis on certain kinds of evaluation, and how we use such evaluations, often narrow in scope, as a measures of "success" and "failure."

Section 3 explores the ways in which gender equity is not viewed as a priority for school reform, and the resulting ultimate bind funding agencies and policymakers find themselves in, even those that are already committed to supporting this issue. I look at the different ways in which funding agencies define and evaluate the impact and importance of this work both locally and nationally and, in doing so, investigate more intensively how foundations and other educational funding agencies frame and define gender in education work. For example, do a majority of educational foundations believe that gender is not an important enough priority to target limited resources for? Do foundations see gender as a synonym for "girls," and thus believe that it is not "inclusive" or "democratic" enough to merit special funding status (Mead, 2001)? Even if foundation staff members recognize that gender issues are worth paying close attention to, what is the impact on the foundation of supporting an issue that the general public and other important stakeholders do not yet recognize? Should foundations try to fund programs that address gender issues without calling attention to the gendered aspects of the programs? What are the advantages and disadvantages of this approach? For example, is there a danger that gender will become subsumed into other educational problems?

Section 4 considers more broadly some of the pitfalls of traditional evaluations when applied to issues of gender and urban education. For example, interviewing or shadowing participants can be very demonstrative, but are rarely cost-effective. Likewise, many funders and policymakers do not value qualitative evaluation at all, believing that it is not scientific or systematic enough.

Section 5 presents some case study examples of alternative approaches to evaluations designed by the Ms. Foundation for Women which are specifically geared toward looking at gender.

Finally, in section 6, I consider how funding organizations decide how and where to distribute scarce resources, recognizing that all girls are not equally disadvantaged, and moreover, that poor and minority *males* are often identified as the group *most at risk* (Bierda, 2000; Connell, 1993; Davis, 2000; Flood and Dorney, 1997; Ogbu and Simmon, 1998). Should foundations reach out to all children equally, or try to concentrate their resources on those groups most in need? How do foundations decide which kinds of gender issues should be their primary focus? What kinds of "proof" do foundations need that their money is being well spent and that their resources are being distributed to those most in need or most worthy? Are the voices of the program participants themselves the most important voices to listen to? What other kinds of outside measures are necessary (for example, test scores, numbers of participants reached, materials generated, et cetera.)? How can evaluations be methodical and intentional, while also being flexible and authentic to those involved? What are the other considerations that drive foundations' decision-making processes? In other words, how relevant, important, or useful is program evaluation at all, in face of political relationships, stakeholder priorities, and long-term funding histories and cultures (Mead, 2001)?

Section 1: Focusing on Gender in Urban Education: Types of Programs and Research

Before I discuss how gender programs are funded, compared, and assessed, I believe it is useful to look briefly at some of the different kinds of programs that fall under the rubrics of "gender and urban education." It is important to note that definitions of terms like *gender equity*, *gender awareness*, *gender bias*, and *gender studies* are not universally agreed upon in education. Nor, for that matter, are the programmatic elements which comprise them. While some define gender equity in urban education as treating boys and girls exactly alike, or at least giving them equal attention, feedback, and resources (Sadker and Sadker, 1994), others advocate teaching to students' (real or perceived, innate or socialized) *differences* (Gilligan, 1982; Gurian, 2003). This may mean that girls are encouraged to work collaboratively while boys are still working competitively; or that girls pay more attention to language arts while boys are strongly encouraged in the fields of math and science, et cetera. Others still see gender equity as a form of affirmative action or remediation, as has become most apparent in attacks on policies such as Title IX. Title IX, designed as an equal opportunity in education law, is most notable for its mandate to assure that girls' sports get the same funding as boys' sports, often necessitating a shift of limited funds and resources from boys to girls.

It is also worth noting that gender equity programs differ considerably in size, as well as in scope, content, length, and goals. Programs geared specifically for teachers in the form of *professional development* or *curriculum development*, usually include a component where participants are asked to engage in self-reflection on their own gender biases, teaching histories, and cultural identities. Shapiro, Sewell and DuCette (1995) write, "We have to realize that who we are greatly affects our thinking about categories such as ethnicity, social class, gender and other areas of difference" (xiii).

Programs developed directly for students themselves take the form of mentoring, building self-esteem, or creating "safe spaces" for students to voice their opinions and discuss different models of masculinity and femininity (for example, The Girls' Action Initiative, The Alice Paul Leadership Center, The Girl Scouts, Girls Inc., et cetera).

More *research-based* programs systematically explore specific questions about classroom/school dynamics such as why girls actually or seemingly "allow" boys to harass them, or why boys are more inclined to take upper level math and science courses. Often referred to as *action research* or *teacher research*, it has been noted by Cohen and Manion (1984) and many others that this type of research is *situational* (concerned with diagnosing a specific problem in a specific context); *collaborative* (teams of researchers and teachers build inquiry communities and work together, bringing diverse experiences and perspectives to the research); and *self-evaluative* (modifications are made continually throughout the project and is thus a process evaluation as well as a product evaluation). Many times students themselves are asked to help develop, conduct, and interpret this research as an integral part of the curriculum itself (Cochran-Smith and Lytle, 1993; Ginsberg, Shapiro, and Brown, 2004). When involving students in research it makes sense to choose topics that are pertinent to issues in their lives, such as sexual harassment, career counseling, self-esteem, social justice, and community activism.

These are just some examples of the kinds of educational programs that could be considered gender-focused. Such programs are still rare, and getting rarer still as the possible punishment inherent in No Child Left Behind (for example, losing federal funds if schools don't improve test scores) become more and more real. As No Child Left Behind appears to be highly concerned with issues of evaluation and accountability, it cannot (and should not) be ignored; however, as the next section of the chapter points out, there are different kinds of assessments, many of which are far more effective and revealing than high-stakes testing.

Section 2: Evaluation and Accountability: Conjoined Twins?

Educational program *evaluation* usually has one (or more) of the following purposes:

- *Assessment.* in order to improve programs in process or to make changes in future programs;
- *Research.* to make comparisons of the relative success of different program models, or to study the impact of programs at different sites and with different constituent groups; and to increase the state of knowledge in the field; and
- *Accountability.* to measure how goals of different stakeholder groups, particularly policymakers and funding agencies, were met (Council on Foundations, 1993).

It is no secret that the third motivation, *accountability*, is often the one that drives program evaluation. Indeed, accountability (Shapiro, Sewell, and DuCette, 1995; Ginsberg, Shapiro, and Brown, 2004) is not only the dominant reason given for program evaluation but plays a significant role in shaping the *ways* that programs are evaluated. Accountability is intricately related to the measures, methods, and tools that are considered legitimate markers of success, as well as the form and content of the final report(s) these evaluations take and who reads them. As Shapiro, Sewell, and DuCette (1995) rightly note, "Assumably . . . all the significant outcomes of education can be objectively measured . . . . implicitly or explicitly, assessment continues to drive the curriculum" (86). Yet they go on to note that "accountability and diversity tend to go in opposite directions" (87) because accountability leads to uniformity and standardization, while diversity leads to a unique curriculum and to the reflection of individual learning styles in assessment techniques (87).

There are some very understandable reasons why both program developers and program funders need to be accountable for what they are doing—not the least of which—because limited and coveted resources are at stake. Nonetheless, it is the premise of this chapter that an overemphasis on accountability has the potential to skew other program goals—such as assessment and research—in ways that can ultimately make program evaluation both less authentic and less useful for everyone involved.

Throughout the chapter I suggest that the current models of school reform and program evaluation, and the traditional markers of "success" (for example, test scores) may not be especially useful when looking at gender-based educational programs. For example, an analysis of students' standardized test scores is unlikely to reveal whether teachers are giving boys and girls equal amounts of attention in the classroom or whether girls are taking greater leadership positions, have increased self-esteem, or are considering a broader array of career options (Ms. Foundation for Women, 2000).

Similarly, the overall "cost-effectiveness" of a program (for example, in terms of its ability to be replicated and numbers of students served) is not necessarily the most important indicator of a worthwhile gender equity program. To be effective, gender programs need to do much more than, as the saying goes, *add women and stir*. Just inserting women into the curriculum is a relatively inexpensive way to make learning seem more equitable, but this method does not address the critical reason *why* women were absent in the first place, and *how* women are usually represented as compared to how men are represented. As Martinez (1995) notes, "[A] numerical increase in textual references and images doesn't promote multiculturalism if the content leaves a fundamentally Eurocentric worldview in place" (101). The same could be said of a patriarchal worldview. And Patrick Finn (1999) reminds us that "if we teach children to critique the world but fail to teach them to act, we instill cynicism and despair" (185). Children are highly aware of bias and inequality. In short, these issues of power and inequity need to be discussed and critiqued; it is not enough to simply try to balance them out in the classroom.

Alternative approaches (discussed at length later in this paper) are often very costly in terms of both financial and human resources and require the sustained commitment inherent in long-term components like mentoring and working closely with parent and community groups. These programs are also not easily replicated because different groups of children, in different schools, with different sets of resources available to them need different kinds of support (AAUW/Research For Action, 1996). When designing any gender-based educational program, it is especially important to take into account the intersections among gender, race, ethnicity, and class. As Ward et al. (2002) notes after conducting a series of focus groups with equity consultants, "Gender Equity initiatives should be specific to and relevant within a context of a child's racial and ethnic community" (4). Ward et al. also found that "[t]eachers were also described as stereotyping students by race, and the charge that teachers hold lower expectations for boys of color was heard across the focus groups" (10).

Evaluators may find it difficult to isolate those students and practitioners who are being affected by gender programs, given the extreme state of flux in urban education. Teachers and administrators have noted that they are continually faced with multiple and competing reform mandates that make it extremely difficult to focus their energies on one particular set of goals (Ginsberg, Shapiro, and Brown, 2004, 147). Moreover, due to heavy dropout rates in urban schools, it is usually not the same group of students that are exposed to constant reforms year after year. With an average student dropout rate of as much as 60 percent at many urban high schools, and a teacher and administrator turnover and vacancy rate that is equally disruptive and alarming, those who participate in these "demonstration" programs are unlikely to be a stable group. For these reasons, gender equity programs (like many other reform programs)—whether professional development programs for teachers or direct support services for students—are unlikely to be good candidates for longitudinal, quantitative evaluation in urban education. It is also worth noting that most gender programs tend to be small "demonstration" projects disconnected from larger school programs and policies (Ward et al. 2002). Thus, when the grant money runs out, the program is quickly forgotten as teachers and administrators are bombarded with new mandates.

In spite of these inherent differences, many funding agencies continue to hold grantees "accountable" to these kinds of measures of success. In fact, the stakes may be even higher for such programs, because of the fact that gender equity remains highly contested and a low priority in most school reform initiatives, and funders and policymakers therefore want to be able to point to immediate and dramatic changes. Unfortunately, this is not an easy task. Many educational stakeholders believe that gender equity is not an important part of educational practice (see section 3) and refuse to prioritize it in any way by giving the participants the needed ongoing support and resources. Thus, the emphasis on accountability has the strong potential to impede the process of change. In the case of one urban reform program designed to reduce dropout rates for inner-city children in New York City, developed by an agency called Cities in Schools (CIS), the evaluator reported that:

> CIS has become burdened with accountability. Reports, meetings, schedules, and agency mandates have taken precedence over children and their needs. All three schools visited had principals extremely

supportive of CIS. But the program is not a part of the school. The teaching and CIS staff are separate and distinct from each other and their attitudes are often competitive and adversarial. (Council on Foundations, 1993, 116)

Moreover, programs which are slow to show change can present a dilemma for foundation staff who fear that "negative" evaluations may "yield information that could reflect negatively on staff judgment" (Council on Foundations, 1993, 17). Although it also should be noted that evaluating grants and programs can potentially provide a positive opportunity for foundations to evaluate their own priorities and practices (Council on Foundations, 1993).

Section 3: Making the Case for Gender Equity in Urban Education

The Issue of Gender in School Reform Work: Not Even on "The List"

In a recent study conducted by the National Center for Schools and Communities (2002), fifty-one diverse community organizations from across the country were given an extensive list of questions regarding their perceived role in local school reform. The study was designed to "identify shared priorities and issues" (1). For the purposes of discussion, an *issue* was defined by the NCSC as "a problem that people understand as being susceptible to policy change and around which they are willing to organize" (1). For example, nearly half the groups surveyed were concerned with supporting (1) after-school enrichment opportunities in their communities; (2) facilitating parent involvement in schools; and (3) increasing funding and accountability.

In its analysis of the interviews, the National Center for Schools and Communities (2002) was able to identify twenty-five "categories" of topics that could be used for intergroup comparisons. Among these twenty-five categories, it was notable that "[g]irls were not to be seen or heard. No interview defined issues, information needs, or context in terms of female students" (13). This finding is especially interesting given that issues of "equity," "safety," and "racism" were among the categories of topics raised by a large number of respondents.

Given the wealth of contemporary educational research citing gender bias and inequities *across* the curriculum, along with high incidence of sexual harassment and gender violence in schools, and the double discrimination faced by poor girls and girls of color, one would think that *girls*, or at the very least *gender*, would appear somewhere on this list of concerns. But they do not. The question is *why not?*

When I posed this question to the Center's executive director, he responded that although he wasn't sure, he suspected that many people still thought of gender issues—such as sexual harassment, or girls' limited enrollment in upper level math and science courses—as an "individual" kind of issue. In other words, they did not see gender as something that needed to be *systematically* and *institutionally* addressed.

Public Perceptions of Gender Bias in Education: Not Even on the Radar Screen

A completely different set of studies conducted by the Frameworks Institute and commissioned by the Caroline and Sigmund Schott Foundation to gauge *public* perceptions of gender equity in education, may offer us some additional clues as to why gender is not seen as a priority in school reform work. Through a series of interviews and focus groups, authors Grady and Aubrun (2000) concluded that gender bias in schools was still a largely invisible issue which did not show up on the American public's radar screen. They furthermore deduced that many Americans still tend to see gender discrimination as a problem in the workplace rather than in the classroom (Grady and Aubrun, 2000).

The Frameworks Institute also uncovered a number of other important findings regarding *how* the general public "frames" and understands the issue of gender equity in education. Embedded in these findings are significant tensions and dilemmas which help to explain why gender equity is often overlooked and undervalued in school reform work. For example, the dominant rhetoric that education is the key to social mobility; that schools treat all children equally; and that school systems are not influenced by larger social, cultural, and political issues is clearly alive in the public's belief that the classroom is an "ideal, controlled environment" (Grady and Aubrun, 2000, 6) where students are protected from rather than exposed to discrimination and bias.

Likewise, the fear that paying closer attention to girls necessarily means taking something away from boys (many of whom are also "at risk") is also evident in the finding that many people resist focusing on the "specific disadvantages faced by girls" (Grady and Aubrun, 2000, 6). In other words, people are likely to resist gender equity if they see equity as a metaphor for remediation or affirmative action. This is most currently obvious in the attacks on Title IX, which many people perceive to be unfairly and unnecessarily draining resources from boys' sports in order to provide more resources for girls.

Indeed, as another important finding suggests, many people believe that teachers actually *favor* girls. This may well be because girls are generally quieter, less disruptive, more compliant than boys, and in many cases, get better grades and are considered to be more "mature" students (AAUW, 1992; Grady and Aubrun, 2000; Orenstein, 1994; Sadker and Sadker, 1994). Yet as these same studies underscore, the end result is often that boys get more attention, and girls get less feedback. Moreover, it is difficult to talk about boys and girls as discrete categories, given that girls living in poverty and girls of color often experience school in extremely different ways than those from middle-class white families and communities. As Orenstein (1994) astutely observed in her ethnographic study of working-class, minority girls at an urban middle school,

> In the classrooms at Audubon, issues of gender are often subsumed by issues of basic humanity, often secondary to enabling a student—any student—to go through the school day without feeling insulted, abused, or wronged by her peers or by her teachers. (137)

It is important to underscore that the word *gender* is often synonymous with *girls*, and therefore that gender equity is somehow a specialized concern that benefits one group over another. Further, there is a concern that these affirmative actions are often without merit, a concern which has become central to educational funding agencies which seek to distribute limited funds in the most democratic and responsible manner. Evaluation of such programs thus has sought to prove not only that the programs themselves are well designed and effective, but that the entire topic is worthy of concern.

Funding for Gender in Education Programs: Addressing "The Bottom Line"

A number of compelling studies underscore that gender is as divisive an issue in the funding world as it is in the school reform and public arenas. In *Gender Matters: Funding Effective Programs for Women and Girls*, for example, Mead (2001) reports,

> The bottom line is that funders have a strong preference for funding so-called universal (or coeducational) programs, and [have] little awareness of the need to consider gender when setting grant making priorities or allocating funds to grantees. (3)

In her study of funders in the Greater Boston area, Mead (2001) identified a number of different rationalizations foundations use to resist focusing on gender. These include: *efficiency*—wanting limited resources to reach the broadest possible audience; *democracy*—wanting programs to be as inclusive as possible; and *relevance*—gender is not the most "critical criteria" for school reform (10–11).

Yet when Mead (2001) studied twenty-five so-called universal coeducational youth development programs for urban teenagers, she found that gender was, in fact, an extremely relevant category. Mead found that these programs did not pay enough attention to the different life experiences of boys and girls, and the ways that these experiences are shaped by gender norms (albeit these "norms" were further shaped by issues of race, ethnicity, and class). For example, Mead notes that women are significantly more likely than men to be living in poverty due primarily to "labor-market segregation and women's significantly greater role in raising children" (17). Thus a program that is concerned with issues of poverty or that seeks to assist poor people must consider these gendered components.

Mead (2001) also notes that because women and girls are socialized differently than men and boys, in mixed gendered groups they may be inclined to talk less, be "reluctant to engage in verbal conflicts" (17), or less likely to take leadership roles. Mead concludes by making a case for more *gender-sensitive* programming rather than *universal gender-blind* programs. Programs may continue to be coed, although Mead suggests that in single-sex programs girls may avoid feeling like the "other," may feel more safe, and may have greater opportunities to exercise leadership abilities. Although Mead is not arguing that gender differences are innate or immutable, her findings underscore the conclusion that "to be effective for women and girls, programs need to take gender into account" (4).

Yet even foundations that have specifically committed to using gender as a guiding focus find this work interrupted by nagging questions of how to achieve their goals in a climate of universality, invisibility, and resistance to using gender as a specific programmatic lens. In the spring of 2000, the Three Guineas Fund (2001) conducted interviews with thirty-one funders and girls' program staff, bringing individuals from each group together for discussion. The resulting report, *Improving Philanthropy for Women and Girls*, provides recommendations for both groups and addresses some of the contradictions and dilemmas also raised in Mead's (2001) research.

The Three Guineas Fund (2001) reports that "[f]oundations often focus on numbers of girls served and cost per girl" (7). Yet program staff dispute that this is the most "effective" measure for evaluation. Instead, program staff advocate for "fewer girls served, smaller staff-to-girl ratios, and more in-depth programming" (7). As one staff member describes it, "Large numbers typically have no long-term impact. When you're reaching 500 people, the impact is superficial" (7). The Three Guineas Fund report also underscores that funders often have unrealistic expectations of evaluation results, expecting change to happen much more quickly than it usually does and that "funders do not often accept qualitative, including anecdotal evaluation measures" (7).

There are, of course, some practical reasons for this. As Mead (2001) rightly notes, "Foundations are both rational and irrational in their decision-making: they are influenced not only by carefully presented research evidence but also by internal and external pressures" (6). In other words, foundations recognize that they need some measure of public and political support for their work, as well as to satisfy board members, donors, and other important stakeholders demands that they are making a real difference and using their resources wisely and productively. In many instances, grantmaking decisions are limited by what Mead describes as a foundation's "history and culture," noting that "prior decisions and standard operating procedures influence and constrain available options and choices in the present" (43).

And it is worth adding that those *seeking funds* are not blind to this reality. As one executive director of a nonprofit candidly told me (when I interviewed him for my dissertation several years ago):

> When you go to big funders, you don't go for one grant. You put yourself up for adoption. You set up a long-term funding relationship. So you can come up with any good idea and can count on them for money. One reason for accountability is not just that public funds are adequately spent, but to maintain the continuity of the relationship with the funder. . . . So the adoption proceeding goes through.

Though it sounds crude, this "adoption process" is no joke for struggling nonprofits that are competing with large numbers of other organizations for an increasingly smaller pool of resources. This ultimately means that program developers and development officers need to design and "sell" programs that can be *easily proven* to be "successful," and thus merit more and future funding. In the current educational climate this means programs that reach large numbers of students and schools, raise test scores, produce "packaged" curriculums, are easily replicable, and are not particularly controversial.

This is in contrast to programs that may "fail" to produce products and raise test scores, yet can succeed in other ways. For example, programs which raise important new questions and insights about how students and teachers are experiencing school; highlight diverse perspectives including formally "silenced" voices; and, perhaps most importantly, teach us about what kinds of educational changes are superficial and what kinds of changes are meaningful and sustainable.

Section 4: Evaluating the Results of Gender Equity Programs in Urban Schools: What Constitutes Success?

Studies such as those summarized above underscore the ways in which gender is still a largely invisible, uncomfortable, contradictory, and misunderstood issue in school reform. Yet this has not stopped a wide variety of organizations—ranging from large urban school districts and state agencies to the smallest nonprofit community groups—from designing programs to address gender inequities and raise awareness of the importance of paying closer attention to gender in schools. Although these programs have received some public attention, most of them are short-lived and have few paper trails.

In a relatively recent literature review of best practices in gender equity and education, commissioned by the Caroline and Sigmund Schott Foundation, Janie Victoria Ward, director of The Alliance on Gender, Culture, and School Practice at Harvard University, and her coauthors noted that it was very difficult to gather information about community-based programs that do gender work in schools because the majority of such gender-based programs operate after school hours and are not aligned with the school's official curriculum, culture, or policies (Ward et al., 2002).

These findings echo those revealed in similar research conducted by the Ms. Foundation for Women (2000) a decade earlier, in which the Foundation sought to "understand what it was about effective programs serving girls and women that made them work," and, in fact, to "prove" that they were working to benefit both girls and their communities (1–2). After conducting an overview of such programs and convening interested stakeholders, the Ms. Foundation concluded that:

> the reality for most girls' programs is that they are not part of an explicit and intentional evaluation process.... Most youth programs do not even have a budget for evaluation, which is typically considered either a luxury item separate from the "real" work of girls' lives or a seemingly meaningless task required by funders. (1)

Yet even when educational programs do have an explicit and comprehensive evaluation plan and budget, these evaluations are often, as the Ms. Foundation suggests, driven by *accountability to funders rather than authentic opportunities for learning*. And the kinds of questions that funders and other stakeholders want answered, such as proof of increased student achievement and sustained changes in school culture, are often difficult to measure, as evaluators are frequently stymied by the reality of life in public (particularly urban) schools.

For example, programs that aim to provide gender-focused professional development for teachers and administrators around gender inequities are often difficult to evaluate because of inconsistent participation and heavy turnover of those practitioners involved. Such was the case with the Gender Awareness Through Education Program (GATE) developed by the Pennsylvania Humanities Council and funded by the Annenberg Foundation, Core States Bank, and the Arco Chemical Company. Without further probing and without considering the larger context in which these programs take place, it would be easy for evaluators to conclude that GATE's low participation levels were due to disinterest in the program. But, as was the case in one such program, low attendance was not simply a reflection on the worth of the program. In all of the schools involved, participants were constantly changing jobs (often in nonlinear ways, such as an art teacher who became the dean of students), retiring, transferring to other schools, or were simply unable to find a common meeting time given the myriad of additional responsibilities each was saddled with. Thus, any sort of quantitative measurements about how many people the program reached and how committed they were to its goals needed to be qualified by qualitative, anecdotal evidence.

Participants in the above mentioned program, for example, reflected upon the value of the program in highly positive terms. One noted that "[i]t's almost as if a consciousness in every word I say and how I present material to my students, even physical eye contact and movement, has changed." Teachers noted repeatedly that one outcome of the project was that they were much more sensitive to their students' viewpoints and perspectives, and more able to engage them in classroom discussions and learning. This is particularly significant given that the schools involved were primarily comprised of poor and minority students—those who research shows are often the most alienated from school, and the most likely to dropout. One teacher noted in his final reflection that, through his participation in the program, "I got insights into the way kids think, their view of the world. This is very important. I know we need to personalize education, understand their thinking."

Teachers also measured success in terms of the kinds of communities which were formed with other educators, and whether or not such communities can/will be sustained after the official program is over. When asked directly, "How do you measure success in reform programs?" one GATE participant responded, "Firstly, by my relationships with other people in the group. Did I keep in touch with anyone in the group, [are we] still in contact?" Another had a similar comment when asked what was the most important aspect of the participating in the program, "Being able to share with other teachers. We get very isolated." Yet another responded, "I think GATE is a very successful program because we're still talking about it" (see Chapter Six of Ginsberg, Shapiro, and Brown, 2004 for more on this topic).

Another problem frequently encountered in the evaluation process concerns how to measure the impact of the program *on students*. In other words, it's all well and good to have more insightful, sensitive teachers, but how do gender programs actually improve student performance and future achievement? Again, in theory this seems like an easy question to answer: Isolate those students involved. Test them and compile data from them at the beginning of the program and at selected intervals throughout. The reality, however, is far harsher. Just as teachers and administrators frequently change positions and responsibilities, the core group of students being affected by a particular program may also be changing constantly. In the program mentioned above, for example, the high school dropout rate averaged 60 percent. In such circumstances it is difficult, if not impossible, to develop any sort of effective "longitudinal study." And those students who did stay at the school over the entire program often had sporadic contact with participating teachers, as evidenced by the comment of one participating teacher who explained in a written reflection:

> That teaching year I had five different groups of students. Four classes that I taught were special education students whom I saw on alternative days for English and History (one day I taught English and the next History for double periods, alternating subjects and days for the schools year). The fifth class I taught was a regular English class that also met every other day for two periods. All these classes were grouped heterogeneously by grade and ability. (From GATE participant's monthly seminar reflection)

The program developers' idea of tracking students longitudinally beyond the pilot phase, while extremely worthwhile, proved to be even more infeasible. It would have been impossible to isolate those students affected only by this particular program and compare them to other groups of students. Perhaps most importantly, the kinds of changes initiated—such as changes in students' self-esteem and career choices—would be extremely difficult to discern from these kinds of multiple choice tests, with so many competing factors and in such a short time frame.

Similar problems are brought to light regarding the issue of documenting "sustained" change in school culture. Most urban schools do not have a single culture; rather different students experience school very differently depending upon their race, class, gender, family support systems, academic ability, which teachers they have, and other factors. Moreover, urban schools are continually in the process of reform. As Hess (1999) suggests,

> Not only are districts pursing an immense number of reforms, they recycle initiatives, constantly modify previous initiatives, and adopt innovative reform A to replace practice B even as another district is adopting B as an innovative reform to replace practice A. (5)

Just as it is difficult to isolate those particular students being affected by a special program, it is equally difficult to isolate the impact of one particular program within the context of a myriad of other changes. The atmosphere of instability that is common in urban schools cannot be taken for granted in the evaluation process. During the course of the program discussed above, for example, the district's superintendent resigned almost immediately after the program began, and a new superintendent was hired bringing an ambitious, district-wide schools reform plan of his own. The new superintendent further made it known, mainly through word of mouth, that he was not in favor of "pilots and demonstration projects." This not only meant that teachers were not rewarded or recognized for the extra time they essentially volunteered to the gender program, but also that they were faced with a multitude of other reforms and expectations—some of which were directly at odds with gender awareness work. This superintendent would leave a number of years later, and the district would eventually be taken over by the state, bringing the entire district into a state of flux and uncertainty.

Ginsberg, Shapiro, and Brown (2004) has suggested that accountability often becomes a question of who to *blame*, rather than how to find workable solutions and create collaborative coalitions. To a certain extent, any accountability system is flawed in that there are many factors that educators and students simply cannot control. These include poverty and racism, as well as the fact that different stakeholders are often working toward different overall goals and objectives. Hargreaves (1994) has spoken of this as the difference between real collegiality and "contrived collegiality."

Although participating teachers' evaluations of the GATE program were highly positive, most indicated a lingering disappointment that the program did not accomplish something more "concrete," something more "replicable." Some participants lamented that they were not able to interest many other teachers and school administrators in the work they were doing around gender, and that while they themselves had changed considerably, the school as an institution remained basically the same. As one teacher said in an end of program interview, "Did it change the school? I would say on a scale of 1–10, maybe about a 3." Another noted similarly, "The discussion died with me at the end of the year. I couldn't communicate to other teachers how to talk about these issues" (from GATE participants monthly reflections).

Program developers and funders expressed a variety of similar concerns about its ultimate "payoff." Some comments in this regard included, "I don't have a sense of how much was got out [of it]. I don't have a measure of translation to the classroom," and "[w]e need experiments that can be more easily translated into replicable programs." The underlying messages, commonly heard in education and evaluation circles, are as follows:

- The most important role of evaluation is to *measure* the end product, as opposed to raising new insights and questions about important and complex issues, providing a forum for educators to discuss and debate; and

- Programs that are not easily packaged and replicated are not worthwhile investments for funders.

Thus, the question remains: What would a gender program evaluation look like where success was measured by the amount of reflection, inquiry, discussion, and genuine learning rather than simply "The Bottom Line"?

Section 5: New Models of Evaluation, New Definitions of "Success," New Juxtapositions of Qualitative and Quantitative Data

There is, of course, a rich history of qualitative evaluation, participatory evaluation, ethnographic evaluation, and teacher/action research (Anderson, Herr, and Nihlen, 1994; Cochran-Smith and Lytle, 1993; Shapiro, Parssinen, and Brown, 1992) which often stands in stark contrast to evaluation based only on "statistics" and "test scores." Although programs which address "women's" issues have become more plentiful since the Ms. Foundation first began such funding in the early 1980s, there still isn't an extensive body of evaluation literature that focuses explicitly on *gender issues* in schools—especially in ways that do not simplistically set boys and girls up in opposition to each other (Skelton, 2001).

This last point is particularly important, because, as Skelton (2001) has emphasized, "[T]he complexity of gender and achievement cannot simply be 'read off' crude, basic data" (165). In considering the experiences of girls *and* boys in schools, one must take a "relational" approach as opposed to an "essentialist" approach. A relational approach understands that boys and girls construct their own cultural identities, based on differences in race, class, ethnicity, and other differences, and that these identities are not fixed but rather different aspects are more prominent in different contexts. The "essentialist" notion that "boys are boys" (Gurian, 2003) does not work in practice. Educators need to consider complex examples of mascul*inities* and femin*inities*, meaning that (1) boys and girls are studied in connection and interaction with each other, not as static beings, but as people who are always interacting with their environment; (2) issues of race, class, ethnicity, sexual identity, and religion are carefully considered as the data is collected and disaggregated;

and (3) ideally the students themselves help to collect and analyze data based on their own (ever-changing) perspectives and experiences. This stands in stark contrast to both the testing approach and also to a hegemonic view which does not consider issues of socialization and power differences.

The question of what an alternative gender evaluation might look like was tackled by the Ms. Foundation for Women (2000) when they convened a group of like-minded foundations and donors to form the Collaborative Fund for Healthy Girls/Healthy Women. The Ms. Foundation set about creating a special fund for the development and support of programs to increase girls' self-esteem, leadership, community activism, and achievement in education. The Collaborative created a $4 million fund to "provide resources over three years to new and existing organizations with programs focused on girls' empowerment and activism" (2). The extremely diverse groups that ultimately received funding included the Asian and Pacific Islanders for Reproductive Health (Long Beach, CA); the Center for Anti-Violence Education (Brooklyn, NY); Mi Casa Resource Center for Women, Inc. (Denver, CO); and Native Action (Lame Deer, MT), among others.

In March 1999, the Collaborative organized the Young Women's Action Team (YWAT), a representative group of girls drawn from six of the grantee programs, who worked alongside young women scholars to develop research questions and evaluation tools for the entire Collaborative. A central question that emerged from this group was, *"How does being in a girl-centered program impact girls' lives?"* This was a question that they wanted to answer on a number of different levels: (1) the individual level; (2) the social network level; (3) the community level; and (4) the institutional level. In other words, the focus was not exclusively on academic achievement as we have come to view it through the lens of standardized testing. While individual achievement was an important outcome to be evaluated, the group also wanted to understand what made girls leaders and what enabled them to pursue social justice policy and programs for the betterment of the entire communities and schools.

Using a *participatory evaluation research approach*, in which the girls themselves played a central and ongoing role in evaluating the programs they participated in, several exciting new evaluation tools were developed and tested. Two of the most interesting, described herein, include the Voice, Action, Comportment, and Opportunity Checklist (VACO) and The International Storytelling Measure (ISM).

These were discrete, short-term programs that mostly took place outside of regular school hours, and that included a relatively steady group of girl participants and adults. These evaluations did not address the question of how to do similar work within the context of schools where just about the only thing that stays steady from year to year is the cafeteria food. Nonetheless, these evaluation tools may be widely considered as examples of evaluations that do more than measure superficial changes, and are, in and amongst themselves, opportunities for learning and reflection.

VACO

VACO was developed as a method of measuring and describing "incremental change in girls' leadership skills and qualities" (Ms. Foundation for Women, 2000). The Ms. Foundation for Women (2000) defines VACO as

(V) Voice: girls' ability to speak on their own behalf
(A) Action: girls' ability to use their voices to act on behalf of themselves and others
(C) Comportment: girls' ability to carry themselves with pride, respect and dignity
(O) Opportunity: girls' ability to ask for and take advantage of new changes and experiences. (1)

The VACO evaluation includes both a pretest and a post-test, although its primary purpose is to *chronicle girls' development as it happens day-to-day*. Staff members pick a minimum of six girls to observe during program activities, taking notes on the different ways that each girl uses her voice and interacts with others. At the end of each observation period, staff members complete a VACO checklist for each of the four VACO categories being observed, noting the existence of certain behaviors such as:

- *Voice.* Challenged another girls' opinion; stated and defended a point of view or idea; expressed analysis of injustice, discrimination, or prejudice; struggled to say something hard about herself in a group.
- *Action.* Organized others to engage in activities without being told; stopped conflict between other girls; resisted pressure from others to go along with something.

- *Comportment.* Looked directly at others; paid attention to the facilitator; listened to peers in the group.
- *Opportunity.* Suggested ways to find more resources; volunteered to do something that is new or challenging; asked to have more responsibility.

The developers of VACO stress that the evaluation, though clearly not quantitative in nature, can provide "statistical documentation" of girls' development as leaders within the various programs they participate in. For this to work, however, certain guidelines must be followed. These include observing the same girls at different points in time and ensuring that the same staff member observes the same girls each time to maintain consistency, among other guidelines. They also stress that VACO should be used with girls in the first week that they begin attending the program in order to produce a clear measure of change.

What makes VACO especially exciting, however, is that the girls themselves are a critical part of collecting and triangulating (for example, cross referencing) the data. For example, at the end of each observation period, staff *and girls* separately complete the same checklist. Not only does this provide a way for staff and girls to compare their observations (thus triangulating data and challenging discrepancies in language and awareness), *but it also provides an excellent tool for the girls' own self-reflection.*

What would it mean for an approach such as VACO to be incorporated into a school's portfolio of assessment tools, to be integrated into the classroom on an ongoing basis, much like standardized tests? The answer to this question is premised on the idea that actions can be as meaningful as "words," and further, that knowing a lot of "facts" does not necessarily translate into important skills like imagination, responsibility, organization, curiosity, risk taking, open-mindedness, and leadership. Thus, as we evaluate students' achievement in school—whether they be boys or girls—we need to consider evaluation tools that "measure" these skills, and furthermore, that ask students to assess themselves as a means for further development. Clearly this is a much more subjective and time-consuming project than testing, but is nonetheless, critically important if we really want to understand how to empower youth to be social leaders of the future.

ISM

Another Ms.-developed evaluation, ISM, stands for the Intentional Storytelling Measure. In some ways, this is a kind of standardized "test," though certainly not one in which there is only one "right" answer. The exercise, which involves girls reading and responding to a number of hypothetical problems, is designed to "see whether girls perceive themselves as capable of acting as agents for change in relationship to their peers, families and communities" (*Ms. Foundation 2000 New Girls Movement: Assessment Tools for Youth Programs* [internal publication]). Students are asked to brainstorm a number of possible solutions, to choose the one that they think is "ideal," and to describe the one which they think they would most "realistically" pursue. As the girls come back to the same stories over time, the ISM helps answer the question, *What is the effect on girls and on their communities of their involvement in social change work (for example, organizing, community service, policy advocacy, and community activism)?*

The stories themselves vary widely. One story focuses on a group of girls who notice that their friend Tina is being physically abused by her boyfriend. As the situation gets more out of control, Tina tells her friends that "it's no big deal," and to "mind their own business." Another story chronicles a group of girls who, after passing an empty lot filled with junk everyday in a neighborhood with no parks or playgroups, begin to think of ways that they could change the situation. A third story concerns a group of girls who go to a school board meeting to make a presentation about developing a new program in the school addressing sexual harassment. The girls are subsequently sexualized and belittled by the male head of the school board who says, "Thank you, beautiful girls, for providing us with such a treat" and then gives them advice about "looking for boyfriends."

As previously noted, after reading each of the stories, the girls are asked to brainstorm possible actions and solutions, including those that are most ideal, and those that are most realistic. The girls then come back to the same stories later in the program, with the ultimate goal of seeing in what ways their responses changed as a result of their participation in the program. It is cautioned that the *same stories* must be used for pre- and posttesting, as different stories are not equivalent. It is also cautioned that when brainstorming, girls should not be prompted toward particular solutions.

At the end of the program, the girls' responses to the stories are carefully coded according to predetermined guidelines, depending upon what the program is trying to accomplish (for example, the skills or qualities that it is hoped the program will impart to girls). The Collaborative suggests that, for the sake of consistency, it is useful to have two people code the same data to determine inter-rater reliability.

Students need to be able to make what are often difficult decisions about how to solve problems for which *there is clearly more than one right answer*. This means considering not only the ideal solution, but those that are most realistic, and those that are most ethically compelling for them personally. Moreover, this evaluation tool stresses the need for evaluation to be an ongoing process where students have the ability to change their answers without suffering penalties.

VACO and ISM are just two examples of an entire package of tools developed by the Collaborative. As noted above, these evaluation tools would need to be reconstructed if they were going to work within a larger and more chaotic school-based environment. Nonetheless, the key components inherent in each are critical. These include:

- The idea that the participants (for example, girls) themselves should be intricately involved in the evaluation process;
- The importance of consistent and intentional gathering of evaluation data, as well as the triangulation of data; and,
- Addressing and evaluating the real issues the programs are trying to address, and the real contexts in which the programs are taking place. In other words, the evaluation should serve a greater purpose than simply proving the successes or failures of a program, but should serve as an ongoing and meaningful tool for self-reflection, problem solving, and relationship building.

This approach may be termed "feminist assessment," which Shapiro (1995) defines as a form of assessment which "assumes questioning is expected regarding all forms of assessment and evaluation of their ultimate uses" (98). In other words, assessment is part of an ongoing process of inquiry, through which learning generates new questions rather than simplistic answers.

While this process may appear to some critics to be too subjective to count as a real evaluation, as the Ms. Foundation for Women (2000) stresses, "[S]cientific documentation and participatory self-reflection do not have to be at odds with each other" (27). The important thing is to make the evaluation process an authentic one for those involved. According to Ms. and the Collaborative,

> Evaluation research does not have to be academic, formulaic, or bureaucratic. Rather, it can be fun and engaging even as it legitimates and empowers our work. Evaluation research can provide us with the real and powerful results of working with young people to change the world. It is also a tool we can make our own and translate into a language that bridges the gaps between age, culture, and experience. (27)

Even so, there are legitimate questions as to how such approaches will be received in an educational climate that is obsessed with numbers (for example, numbers of participants served, numbers of dollars expended, numbers of test scores, et cetera). Even those foundations and other funding agency staff that are interested in an alternative method of assessment of gender-based school reform must give practical consideration to the stakeholders who oversee and judge their work and decision making.

Section 6: Resurfacing Questions for Program Funders and other Stakeholders

This paper, while grounded in contemporary educational research, also reflects upon my own experiences as a program officer at a non-profit educational organization, as the director of a professional development program piloted in a large urban school district, as an evaluator of many different gender equity initiatives, and as a consultant to several national foundations seeking to support programs that address gender issues in schools. Indeed, the impetus for this work comes from attending many roundtables, focus groups, and advisory meetings comprised of educational and gender experts where I found that the same compelling questions kept resurfacing.

Just as program developers struggle with the language they use to describe their programs, most foundations also spend a considerable amount of time designing mission statements, funding goals, and request for proposals. Those foundations that want to help girls in particular, struggle with whether they should state this overtly in their mission statement and request for proposals. The alternative, to use more neutral language like "gender" (which is more inclusive of boys) or even better "diversity" (which is inclusive of everyone), is more likely to receive widespread public support and avoid criticism that the foundation is favoring one group over another.

On the positive side of switching to language from girls to gender or diversity is the recognition that biological sex is not the only issue at stake. Rather, the focus here is on the construction of norms and roles for girls and boys, and the different life experiences and opportunities that result from them. The negative side of this language shift is that, as Mead (2001) suggests, programs that don't specifically target girls threaten to subsume their needs, voices, and perspectives, within a broader more universal (for example, white male) framework.

Another related question that foundations struggle with is what is meant by the term *equity*? Of course, once one starts down this road, the inevitable questions regarding the relationship among gender, race, class, ethnicity, and other differences arise. When we talk about girls as a "disadvantaged group" in need of special programs and consideration, this is, of course, a relative term. Not all girls are disadvantaged in the same way. Obviously girls of color and poor girls face more complex and daunting challenges than middle-class white girls. Many go as far as to argue, with good reason, that *African American boys* are actually the most at-risk group. As one participant in a professional development program designed to raise awareness about gender inequities admitted in an open-ended written reflection:

> An early struggle at my setting . . . was that many staff members believe that African-American males are so deeply at-risk that it is superfluous and academic elitism to devote professional development time to exploring sexism in our classrooms.

Indeed, in a series of recent interviews I conducted with education, gender, and policy experts across the country—such as presidents of schools of education and research institutes, executive directors of foundations and women's funds, and leaders of political advocacy organizations—a significant number felt that gender issues paled in comparison to the much more tangible tragedies of racism and poverty. These leaders persistently pointed to the abundance of African American and Latino youth in poor urban communities with no positive role models, inadequate school funding, decrepit and unsafe school buildings, large numbers of uncertified teachers, and alarming dropout rates.

Of course, it is important to point out that at least 50 percent of these children referred to above are *girls*, and, moreover, that boys in these communities are also damaged and constrained by unhealthy and rigid definitions of masculinity, et cetera. Nonetheless, the pervasive belief was that gender was simply not the dominant concern. This leads many foundation board members and staff members to question how they can support change and build broad-based advocacy around what many people feel is a low priority issue.

The following questions are raised repeatedly:

- Should we start where people (for example, politicians, administrators, funders, teachers, parents, et cetera) *are* in their thinking and understanding of the issues, or where we *want them to be?*

- If girls and gender remain the focus of the work, how much attention (if any) should one pay to gender in isolation from other forms of difference?
- In order to adequately and genuinely address issues of race and class should different groups of girls (or boys) be targeted as higher priorities for funding?
- Should we try to address all these issues at once and cater to all groups in need, or will we end up simply diluting our resources to the point in which we are effective for no one?

These are critically important and very difficult questions to answer.

Another series of questions frequently raised in these meetings concerns whether to *build on existing work* or *create new work*. The continuing impetus to create new programs reflects not only a lack of awareness about the important work already being piloted by women's organizations, community groups, school districts, and other reform agencies across the country, but an impatience to see concrete and large-scale changes as soon as possible. In "A Conversation About Girls" (Valentine Foundation, 1990), a series of conversations among funders, researchers, and gender equity advocates, organized by the Valentine Foundation and Women's Way, it is stressed that "[m]any programs labeled a failure have simply not been funded long enough for the girls to achieve the program goals. This is a financing problem, not one of program design or inability of girls to respond" (4). This dilemma thus circles back to the fundamental issue of this essay: gender and program evaluation.

- What kinds of measures of change do we look for and recognize as important and valid?
- How quickly do we need to see documentation of these changes?
- Who is the primary audience for these evaluations?

Since funding agencies are faced with the difficult question of how they will evaluate the impact of the programs they fund and make decisions about future funding based on internal and external evaluations, many program directors fear that any negative evaluation will automatically kill future funding of their programs. It is thus not uncommon for them to reduce the complexity and sophistication of the evaluation tools to ones in which only the positive (and universally recognized) outcomes are highlighted. Other programs do not have the staff or the know-how to conduct a formal evaluation of their work and are unable to find the necessary support (financial and human) to make this work a priority (Council on Foundations, 1993). And even for those groups that are developing and using cutting-edge evaluation tools, such as the Collaborative Fund for Healthy Girls/ Healthy Women, it is not yet clear how these kinds of evaluations will hold up when push comes to shove (for example, when large sums of money, resources, and political goodwill are at stake).

This is one of the reasons why, when we think about the relationship between funding and evaluation, we need to think beyond the traditional models of philanthropy. In *A Conversation about Girls* (Valentine Foundation, 1990), the authors stress that one of the challenges for foundations supporting this kind of work is learning to work collaboratively. They also note that it is important to "include the young girls [or boys] themselves within the process of planning and designing programs," further underscoring that "[w]e must not let ageism interfere with the design of programs that may be more effective than those we ourselves design" (7). An example of a foundation acknowledging such work is the Michigan Women's Foundation which sponsors young girls, training them in philanthropy, and then has them develop RFPs (request for proposals) and awards grants equaling $20,000 per year.

The *Conversation about Girls* (Valentine Foundation, 1990) raises another important point:

"The resources of private foundations alone, however, are not sufficient. The larger task lies with legislative bodies that must exercise corporate responsibility, addressing in concrete programmatic and financial terms problems that affect the life and health of the entire society" (6). In other words, while focusing on gender equity in urban education and while making sure this work genuinely includes boys as well as girls, we need to look beyond schools themselves, understanding that this is a social justice issue, and not a passing educational fad. Funding agencies interested in supporting equity programs are constantly being asked to explain and defend why such programs are important, and how they can make a real difference in real students' lives and in their urban communities as well. This is most especially difficult when the pressure is to focus on the three R's as early as preschool, preparing children at younger and younger ages to pass the narrowly designed tests which have become so central to our entire educational system.

Finally, it is useful to point out and remember that addressing gender in education means bringing a variety of stakeholders to the table, not just in the early planning stages or as underutilized "advisors" but as real participants throughout the entire program;

paying attention to *process* as well as *product*, meaning that teaching and learning go beyond memorizing facts, and that teachers and students have complex relationships which make gender study messier but significantly richer; and always having larger goals in mind (for example, moving beyond the "add women and stir" approach toward one where gender becomes an ongoing critical lens of analysis, as does race, class, ethnicity, and other differences in human identity and experience). The Council on Foundations (1993) further adds some very useful suggestions to this list, including that participants (1) make evaluation planning an integral part of the program itself, starting as early as possible; (2) make sure evaluation plans have built-in flexibility which will allows for change if the project moves in another direction or the original evaluation plan is not collecting useful data; (3) respect previous work, and, whenever possible, build on existing knowledge; and (4) try to bring like-minded funders together to pool funds, and to share ideas and information (252–265).

Despite not making the official list of reform priorities, showing up on the public radar screen, or addressing the "bottom line," the very fact that these questions are continuing to be raised and taken seriously provides a measure of hope that the issue of gender equity in education will not simply "go away."

Note

This chapter was originally published in *Perspectives on Urban Education* 3 (2) in 2005.

References and Further Reading

AAUW (American Association of University Women). (1992). *How schools shortchange girls: A study on major findings on girls and education.* Washington, DC: Educational Foundation and National Education Association.

AAUW (American Association of University Women)/Research for Action. (1996). *Girls in the middle: Working to succeed in school.* AAUW Report. Washington, DC: AAUW.

Anderson, G., Herr, K., and Nihlen, A. S. (1994). *Studying your own school.* Thousand Oaks Press, CA: Corwin.

Bierda, M. (2000). The myth of the African American male. *WEEA Digest*, 9 (10).

Cochran-Smith, M. and Lytle, S. (1993). Inside/Outside: Teacher Research and Knowledge. New York, NY: Teachers College Press.

Cohen L., and Manion, L. (1984). Action research. In J. Bell, T. Bush, A. Fox, J. Goodey, and S. Goulding (Eds.), *Conducting small-scale investigations in educational management* (41–71). London: Harper & Row.

Connell, R. W. (1993). Disruptions: Improper masculinities and schooling. In L. Weiss and M. Fine (Eds.), *Beyond silenced voices: Class, race, and gender in United States schools* (191–208). New York: State University of New York Press.

Council on Foundations. (1993). *Evaluation for foundations: Concepts, cases, guidelines and resources.* San Francisco: Jossey-Bass.

Davis, J. E. (2000). Mothering for manhood: The (re)production of a black son's gendered self. In C. Brown and J. Davis (Eds.), *Black sons to mothers: Complements, critiques and challenges for cultural workers in education* (51–70). New York: Peter Lang.

Finn, P. (1999). *Literacy with an attitude: Educating working class children in their own self interest.* Albany: State University of New York Press.

Flood, C., and Dorney, J. (1997). Breaking gender silences in the curriculum: A retreat intervention with middle school teachers. *Educational Action Research* 5(1): 71–86.

Francis, B. (2000). *Boys and girls and achievement: Addressing classroom issues.* London: RoutledgeFalmer.

Gilligan, C. (1982). *In a Different Voice: Psychological Theory and Women's Development.* Cambridge, MA: Harvard University Press.

Ginsberg, A. (1999). *When policymakers and practitioners partner: A stakeholder analysis of an urban school reform program.* Philadelphia: University of Pennsylvania, Graduate School of Education.

Ginsberg, A., Shapiro, J. P., and Brown, S. P. (2004). *Gender in urban education: Strategies for student achievement.* Portsmouth, NY: Heinemann Press.

Grady, J., and Aubrun, A. (2000). *Talking gender equity and education: A FrameWorks message memo.* FrameWorks Institute. Report commissioned by the Caroline and Sigmund Schott Foundation, Cambridge, MA.

Gurian, M. (2003) *Boys and Girls Learn Differently: An Action Guide For Teachers.* San Francisco, CA: Jossey-Bass.

Hall, P. M. (1997). "Epilogue: Schooling, gender, equity and policy." In B. J. Bank and P. M. Hall (Eds.), *Gender equity and schooling.* New York: Garland Publishing.

Hargreaves, A. (1994). *Changing teachers, changing times: Teachers' work and culture in the post-modern age.* New York: Teachers College Press.

Hess, F. (1999). *Spinning wheels: The politics of urban school reform.* Washington, DC: Brookings Institution Press.

Leadbeater, B., and Way, N. (1996). *Urban girls: Resisting stereotypes, creating identities.* New York: New York University Press.

National Center for Schools and Communities. (2002). *Unlocking the schoolhouse door: The community struggle for a say in our children's education.* New York: Fordham University.

Martinez, E. (1995). Distorting Latino history: The California textbook controversy. In Levine, Lowe, Peterson, and Tenorio (Eds.), *Rethinking Schools: An Agenda for Change* (100–108). New York: New Press.

Mead, M. (2001). *Gender matters: Funding effective programs for women and girls.* Report. Medford, MA: Tufts University.

Ms. Foundation for Women. (2000). *The new girls' movement: New assessment tools for youth programs.* Report and evaluation tool kit. New York: Ms. Foundation for Women.

National Center for Schools and Communities. (2002). *Unlocking the school house door: The community struggle for a say in our children's education.* Report. New York: Fordham University.

Ogbu, J. (1987). *Minority education and caste: The American system in cross-cultural perspective.* New York, NY: Academic Press.

Ogbu, J., and Simmon, H. (1998). Voluntary and involuntary minorities: A cultural-ecological theory of school performance with some implications for education. *Anthropology of Education Quarterly* 29(2).

Orenstein, P. (1994). *School girls: Young women, self-esteem and the confidence gap.* New York: Doubleday.

Research for Action. (1996). *Girls in the middle: Working to succeed in school.* Washington, DC: American Association of University Women Educational Foundation.

Sadker, M., and Sadker, D. (1994). *Failing at fairness: How our schools cheat girls.* New York: A Touchstone Book.

Shapiro, J. P., Parssinen, C., and Brown, S. (1992). Teacher-Scholars: An action research study of a collaborative feminist scholarship colloquium between schools and universities. *Teacher and Teacher Education* 8(1): 91–104.

Shapiro, J. P., Sewell, T. E., and DuCette, J. P. (1995). *Reframing diversity in education.* Lancaster: Technomic Publishing Company.

Skelton, C. (2001). Typical boys?: Theorizing masculinity educational settings. In B. Francis and C. Skelton (Eds.), *Investigating gender: Contemporary perspectives in education* (164–176). Philadelphia, PA: Open University Press.

Three Guineas Fund. (2001). *Improving philanthropy for girls' programs.* Report. San Francisco, CA: Three Guineas Fund.

Ward, J. V., Rothenberg, M., Benjamin, B. C., and Feigenberg, L. (2002). *Results from focus groups and interviews with gender equity consultants.* Report commissioned by the Caroline and Sigmund Schott Foundation, Cambridge, MA.

Valentine Foundation. (1990). *A conversation about girls.* Conference Proceedings. Philadelphia, PA.

Notes on Contributors

Axel Aubrun is a Senior Research Fellow at the FrameWorks Institute, with a background in psychological anthropology. His research takes an interdisciplinary approach to problems of communication and motivation, integrating perspectives from fields such as cognitive anthropology and evolutionary psychology. Previous public policy publications include "Reducing Violence: Cultural Models and Conceptual Metaphors that Encourage Violence Prevention" (coauthored with Joe Grady, for the Benton Foundation). Prior to joining FrameWorks, Aubrun was a lecturer in cultural anthropology at the University of California at San Diego, and Public Relations Manager at an advertising firm in San Diego.

P. Catlin Fullwood of On Time Associates has been Girl's Best Friend Foundation's evaluation trainer since 2000. Her expertise in participatory evaluation research and how it can be used with and by youth is endless. Her knowledge of work with girls and young women is longstanding. She also serves as a mentor to younger consultants.

Marybeth Gasman is an Assistant Professor of Higher Education in the Graduate School of Education at the University of Pennsylvania. Her work focuses on the history of philanthropy and African Americans. Her most recent book is *Envisioning Black Colleges: A History of the United Negro College Fund* (Baltimore: Johns Hopkins University Press, 2007). She also the editor of *Uplifting a People: African American Philanthropy and Education* (with Katherine Sedgwick).

The Girl's Best Friend Foundation was established to promote and protect the human rights of girls and young women by advancing and sustaining policies and programs that ensure their self-determination, power, and well-being. The foundation supports those who challenge the status quo by offering alternatives to the societal messages that girls and young women receive. This foundation is in the process of closing. They will make no grants after December 2007. When Cyndie McLachlan founded GBF in 1994, it was with the plan to be a short-term foundation, not a perpetual one.

Alice Ginsberg is an independent consultant who specializes in working with foundations and nonprofits on educational issues such as gender, equity, cultural diversity, and urban school reform. She worked for eight years as a program officer at the Pennsylvania Humanities Council (PHC), the state's partner to the National Endowment for the Humanities. She has developed and evaluated programs widely for foundations such as the Ms. Foundation for Women, The Schott Foundation, Lucent Technologies, The Annenberg Foundation, The Howard Heinz Foundation, among others. She is the first author of *Gender in Urban Education: Strategies for Student Achievement* (Portsmouth, NH: Heinemann, 2004). Dr. Ginsberg was also the Director of the Gender Awareness Through Education (GATE), a three-year professional development program for teachers, administrators, and parents at four urban public schools.

Joseph Grady is a Professor of Linguistics at the University of Maryland, Department of English, currently working with the FrameWorks Institute. His academic research and publications focus on the relationship between metaphor and other aspects of thought and communication. He has conducted a number of studies related to public policy, including "Why Early Ed Benefits All of Us: Communications Strategies For Advocates of Early Childhood Education" (coauthored with George Lakoff, for the Benton Foundation). In addition, Grady spent a number of years as a consultant, applying the principles of linguistics to developing brand names.

GrantCraft, a project of the Ford Foundation, makes available a series of publications and videos designed to prompt discussions among foundation practitioners about strategic and tactical lessons in philanthropy. The publications and videos are based on the experience of grantmakers at the Ford Foundation and a wide range of other funders and grantees.

Margaret Hempel is the Vice President of Programs at the Ms. Foundation. She is responsible for managing the foundation's grantmaking and training programs that ensure low-income women and women of color have the resources to engage in advocacy and organizing efforts in state, tribal, and national arenas. This includes work on issues ranging from health and safety to economic security and civic engagement. Before coming to the Ms. Foundation, Ms. Hempel spent nine years at the Ford Foundation in the Human Development and Reproductive Health Unit, first as a program officer and then as Deputy Director. There, she helped lead the design, implementation, and evaluation of a worldwide reproductive health program, making grants in the areas of public policy, sexuality education, media, and the impact of welfare reform on reproductive health and rights.

The Michigan Women's Foundation is dedicated to the full participation of women and girls in society. The foundation not only funds projects, but creates, designs, and launches innovative programs. **Young Women for Change** is a group of young women ages 14–18, of diverse backgrounds, who assess the needs of local girls/young women and grant funds to nonprofit organizations working to serve girls/young women in their local communities. Young Women for Change is a program of the Michigan Women's Foundation.

Shirley Mark is Director of the Lincoln Filene Center for Community Partnerships at the Tufts University College of Citizenship and Public Service. The Lincoln Filene Center is focused on creating opportunities for the Tufts community to be civically engaged in local schools and nonprofit organizations as well as supporting the capacity needs of partner organizations. Prior to Tufts, Shirley worked in philanthropy as a grantmaker and consultant. At the Caroline and Sigmund Schott Foundation, Shirley was Program Director and developed grants initiatives to promote gender equity in urban public schools. She has consulted with the Boston Foundation and Hyams Foundation and has served on numerous local and national grantmaker committees, including the United Way of Massachusetts Bay, Asian Americans and Pacific Islanders in Philanthropy, Boston Funders for Women and Girls, Grantmakers for Education, and others.

Molly Mead is the Lincoln Filene Professor of Citizenship and Public Service at Tufts University. She works with faculty, students, and community residents to infuse education for active citizenship throughout the university. She is also a faculty member in the Department of Urban and Environmental Policy at Tufts University where she has taught gender and public policy, leadership and the field work course. Her 1994 research report, "Worlds Apart: Missed Opportunities to Help Women and Girls," helped us understand why foundations generally give such little funding to programs for women and girls. She has worked since then to increase knowledge about the important role these programs play in serving women and girls. In her book from MIT Press, *Effective Philanthropy*, she makes the case for funding women and girls as an essential aspect of successful grantmaking.

Ami Nagle is a research, evaluation, and planning consultant. Ms. Nagle works with charitable foundations, nonprofit organizations, and government in the areas of early childhood, health, economic security, K-12 education, and community development.

Jessica Palmert serves as the Girl World Coordinator at Alternatives, Inc. a not-for-profit youth and family services agency.

Allyson Pimentel is a graduate of Harvard University's program in Human Development and Psychology. She currently works as a researcher in the Social Personality Program at the City University of New York and a psychologist at NYU-Bellevue Hospital Center. Her research and clinical work focus on gender issues in ethnic minority communities and psychological development across the lifespan.

Michael Reichert has been a practicing child and family psychologist for more than 20 years. He created and directed for 10 years a youth antiviolence program sponsored by Philadelphia Physicians for Social Responsibility and is the coauthor with Sandra Boom, M.D., of *Bearing Witness: Violence and Collective Responsibility*. More recent publications include chapters in the books *Construction Sites: Excavating Class, Race, and Gender Among Urban Youth* (Lessons From A Boys' School) and *What About the Boys?* (Rethinking Masculinities: New Ideas for Schooling Boys).

Rosa Smith served as a school superintendent in Columbus, Ohio and Beloit, Wisconsin prior to joining the Schott Foundation as executive director. She also served as assistant superintendent, high school principal and teacher in Minneapolis and Saint Paul, Minnesota, and South Bend, Indiana, respectively. Smith's current civic and professional leadership roles include National Association for the Education of African American Children with Learning Disabilities Professional Advisory Council, Bessie Tartt Wilson Children's Foundation Board of Directors, The Home for Little Wonderers Board of Directors, Annenberg Institute Special Education & Systems Reform Working Group, National Fund for Research on Black Philanthropy, Association of Black Foundation Executives Research Advisory Group, The Twenty-First Century Foundation Black Men and Boys National Resource Center Advisory Committee.

The Valentine Foundation is a Philadelphia area grantmaking organization that provides charitable funds to organizations or programs that empower women and girls. Our grantmaking recognizes and develops the full potential of women and girls and works to change established attitudes that discourage women and girls from recognizing that potential. The Valentine Foundation trustees commit a minimum of half of their grants to programs for girls and the balance to programs for women. The programs for women must include advocacy for social change. As a grantmaking organization, the Valentine Foundation's emphasis is on programs for women and girls that effect fundamental social change—to change attitudes, policies, or social patterns.

Marisha Wignaraja has worked at foundations and nongovernmental organizations in the United States, Sri Lanka, and Switzerland. She brings a background in international human rights, child and youth advocacy, and social change philanthropy to her work. Marisha is currently with the Atlantic Philanthropies in New York. She is responsible for helping U.S.-based advocacy organizations strengthen their reach and capacity to help disadvantaged children and youth. Most recently, Marisha was an independent consultant. She worked with the Ms. Foundation for Women on documenting the lessons learned from the Collaborative Fund for Youth-led Social Change. Marisha's work documented how individual, organizational, and community identity impact youth-led social change efforts. She is coauthor of *Culture and Context: The Collaborative Fund for Youth-Led Social Change."* Prior to her consultancy, Marisha was a program officer at the Ms. Foundation for Women.

Supplementary Bibliography

Activists, governor in New Jersey launch new women's federation. *Responsive Philanthropy* (Fall 1995): 7.
Addelston, J. (1995). "Exploring masculinities: gender enactments in preparatory high schools." Unpublished Thesis (PhD) CUNY.
AAUW (American Association of University Women). (1992). *How schools shortchange girls: A study on major findings on girls and education.* Washington, DC: Educational Foundation and National Education Association.
———. (1992). *AAUW report: How schools shortchange girls.* Washington, DC: American Association of University Women.
American Association of University Women/Research for Action. (1996). *Girls in the middle: Working to succeed in school.* AAUW Report. Washington, DC: AAUW.
Anderson, G., Herr, K., and Nihlen, A. S. (1994). *Studying your own school.* Thousand Oaks, CA: Corwin Press.
Atkins, J. M. (1998). Philanthropy in the mountains: Martha Berry and the early years of the Berry schools. *Georgia Historical Quarterly* 82(4): 856–876.
Aubrun, A., and Grady, J. (2000). *How Americans understand teens: Findings from cognitive interviews.* Washington, DC: Frameworks Institute.
Bane, M. J., and Ellwood, D. T. (1994). *Welfare realities: From rhetoric to reform.* Cambridge, MA: Harvard University Press.
Batts, L. P. (2000). "Female leadership in philanthropy: Perceptions and challenges." Thesis (EdD) University of San Francisco.
Bednall, J. (1995). *Teaching boys to become "gender bi-lingual."* Hunting Valley, OH: University School Press.
Bem, S. L. (1993). *The lenses of gender: Transforming the debate on sexual inequality.* New Haven, CT: Yale University Press.
Bierda, M. (2000). The myth of the African American male. *WEEA Digest* 9 (10).
Bloom, S., and Reichert, M. (1998). *Bearing witness: Trauma and collective responsibility.* Binghamton, NY: Haworth.
Blumberg, R. A. (1991). *Gender, family and economy: The triple overlap.* Newbury Park, CA: Sage Publications.
Bly, R. (1990). *Iron John.* New York: Vintage Books.
Bonavoglia, A. (1989). *Far from done: The status of women and girls in America.* New York: Women and Foundations/Corporate Philanthropy.
———. (1991). *Making a difference: The impact of women in philanthropy.* New York: Women and Foundations/Corporate Philanthropy.
Bonavoglia, A. (1992). *Getting it done: From commitment to action on funding for women and girls.* New York: Women and Foundations/Corporate Philanthropy.

Bordt, L. (1997). *The structure of women's nonprofit organizations.* Bloomington, IN: Indiana University Press.

Bressi, D. E. (1999). "Women and philanthropy: Making a difference in higher education." Thesis (EdD) University of Tennessee.

Broverman, I., Vogel, S. R., Broverman, D. M., Clarkson, F. E., and Rosenkrantz, P. S. (1972). Sex role stereotypes: A current appraisal. *Journal of Social Issues* 29: 59–78.

Browers, R. (1993). "Review of the integration of gender concerns in the work of the DAC. Theme 1 of the assessment of WID policies and programs of DAC members." Prepared for the Operations Review Unit of the Directorate General for International Co-operation of the Netherlands. Institute of Social Studies International Services, The Hague, Netherlands, Mimeo.

Buckingham, J. (1999). *The puzzle of boys' educational decline: A review of the evidence.* Canberra, AU: Center for Independent Studies, www.cis.org.au (accessed May 3, 2001).

Bulhan, H. (1985). Black American's and psychopathology: An overview of research and therapy. *Psychotherapy* 22: 370–378.

Burton, J. D. (1997). Philanthropy and the origin of educational cooperation: Harvard college, the Hopkins trust, and the Cambridge grammar school. *History of Education Quarterly* 37(2): 141–161.

Buvinic, M., Gwin, C., and Bates, L. M. (1996). *Investing in women: Progress and prospects for the World Bank.* Baltimore, MD: Johns Hopkins University Press.

Camino, L. (1995). Understanding intolerance and multiculturalism: A challenge for practitioners, but also for researchers. *Journal of Adolescent Research* 10 (1).

Capek, M. E. S. (1998). Women and philanthropy: Old stereotypes, new challenges. A Monograph Series. Volume One, *Women as donors: Stereotypes, common sense, and challenges*; Volume Two, *Foundation support for women and girls" "Special interest" funding or effective philanthropy?*; Volume Three, *The women's funding movement: Accomplishments and challenges.* Battle Creek, MI: W. K. Kellogg Foundation, http://www. WFNET.ORG.

Capek, M. E. S., and Mead, M. (2007). Funding Norm doesn't fund Norma: Women, girls and philanthropy. In Neil Carson (Ed.), *The state of philanthropy in America.* Washington, DC: National Committee for Responsive Philanthropy.

Capek, M. E. S., and Hallgarth, S. A., with Abzug, R. (1995). *Who benefits, who decides? An agenda for improving philanthropy: The case for women and girls.* New York: National Council for Research on Women.

Capek, M. E. S., and Mead, M. (2006). *Effective philanthropy: Organizational success through deep diversity and gender equality.* Boston, MA: MIT Press.

Carlton-LaNey, I., Hamilton, J., Ruiz, D., and Alexander, S. C. (2001). "Sitting with the Sick": African American women's philanthropy. *Affilia* 16(4)(Winter): 447–466.

Chamberlain, M. K., and Bernstein, A. (1992). Philanthropy and the emergence of women's studies. *Teachers College Record* 93(3): 556–568.

Chakravartty, S. (1991). *Far from done: Women, funding and foundations in North Carolina and the Southeast*. New York: Women and Foundations/Corporate Philanthropy.

———. (1992). *Far from done: Women, funding and foundations in Wisconsin*. New York: Women and Foundations/Corporate Philanthropy.

Children Now. (1999a). *Boys to men entertainment media: Messages about masculinity*. Oakland, CA: Children Now.

———. (1999b). *Boys to men sports media: Messages about masculinity*. Oakland, CA: Children Now.

———. (1999c). *Fair play? Violence, gender, and race in video games*. Oakland, CA: Children Now.

Clark, P. (1995). Risk and resiliency in adolescence: The current status of research on gender differences. *Equity Issues* 1(1).

Clift, E. (Ed.) (2005). *Women, philanthropy, and social change: Visions for a just society*. Hanover, NH: University Press of New England.

Cohen, L., and Manion, L. (1984). Action research. In J. Bell, T. Bush, A. Fox, J. Goodey, and S. Goulding (Eds.), *Conducting small-scale investigations in educational management* (41–71). London: Harper and Row.

Colbert, Ann. (1996). Philanthropy in the newsroom: women's editions of newspapers, 1894–1896. *Journalism History* 22: 10, 90.

Comitini, P. (2005). *Vocational philanthropy and British women's writing, 1790–1810*. Burlington, VT: Ashgate Publishing.

Conlin, M. (1993). The new gender gap. *Business Week*, May 26, 74.

Connell, R. W. (1987). *Gender and power: Society, the person and sexual politics*. Stanford, CA: Stanford University Press.

———. (1989). Cool guys, swots and wimps: The interplay of masculinity and education. *Oxford Review of Education* 15(3): 291–303.

———. (1993). Disruptions: Improper masculinities and schooling. In L. Weiss and M. Fine (Eds.), *Beyond silenced voices: Class, race, and gender in United States schools* (191–208). New York: State University of New York Press.

———. (1995). *Masculinities*. Berkeley: University of California Press.

Connell, R. W., Ashenden, D. J., Kessler, S., and Dowsett, G. W. (1982). *Making the difference: Schools, families, and social division*. Boston: Allen & Unwin.

Cordes, C. (1985). Black males at risk in America. *APA Monitor*, January, 9–10, 27–28.

Council on Foundations. (1993). *Evaluation for foundations: Concepts, cases, guidelines and resources*. San Francisco: Jossey-Bass.

Courtenay, W. H. (2003). Key determinants of the health and well-being of men and boys. *International Journal of Men's Health* 2(3)(January): 1–30.

Crocker, R. (1996). From widow's mite to widow's might: The philanthropy of Margaret Olivia Sage. *American Presbyterians* 74(4): 253–264.

Dance, J. 2002. *Tough fronts: The impact of street culture on schooling*. New York: RoutledgeFalmer.

Danziger, S. H., and Gottschalk, P. (1995). *America unequal*. New York: Russell Sage Foundation.

Danziger, S. H., and Weinberg, D. H. (1986). *Fighting poverty: What works and what doesn't*. Cambridge, MA: Harvard University Press.

Davis, J. E. (2000). Mothering for manhood: The (re)production of a black son's gendered self. In C. Brown and J. Davis (Eds.), *Black sons to mothers: Complements, critiques and challenges for cultural workers in education* (51–70). New York: Peter Lang.

Diaz, W. A. (1996). The behavior of foundations in organizational frame: A case study. *Nonprofit and Voluntary Sector Quarterly* 25(4): 453–469.

Dooley, C., and Fedele, N. M. (2004). Mother and sons: Raising relational boys. In J. V. Jordan, M. Walker, and L. M. Hartling (Eds.), *The complexity of connection* (220–249). New York: Guilford Press.

Dredge, S. C. (2001). "Accommodating feminism: Victorian fiction and the nineteenth-century women's movement (Anne Bronte, Elizabeth Gaskell, Charlotte Bronte, George Eliot)." Thesis (PhD) McGill University (Canada).

Dubbs, P. J., and Whitney, D. D. (1980). *Cultural contexts: Making anthropology personal*. Boston, MA: Allyn & Bacon.

Dzuback, M. A. (1993). Women and social research at Bryn Mawr college, 1915–40. *History of Education Quarterly* 33(4): 579–608.

Eberhart, C. V., and Pratt, J. (1993). *Minnesota philanthropy: Grants and beneficiaries*. St. Paul, MN: Minnesota Council of Nonprofits.

Elium, D., and Elium, J. (1992). *Raising a son*. Hillsboro, OR: Beyond Words Publishing.

Elliot, D. W. (1995). Sarah Scott's "Millennium Hall" and female philanthropy. *Studies in English Literature, 1500–1900* 35: 535.

Ellwood, D. T. (1988). *Poor support: Poverty in the American family*. New York: Basic Books.

Erkut, E., Fields, J. P., Sing, R., and Marx, F. (1996). Diversity in girls' experiences: Feeling good about who you are. In B. J. Ross Leadbeater and N. Way (Eds.), *Urban girls: Resisting stereotypes, creating identities* (53–64). New York: New York University Press.

Faludi, S. (1994). The naked Citadel. *New Yorker*, September 5, 62.

Farrell, W. (1993). *The myth of male power*. New York: Simon & Shuster.

Feminist Majority Foundation. (1991). *Empowering women in philanthropy*. Arlington, VA: Feminist Majority Foundation.

Finn, P. (1999). *Literacy with an attitude: Educating working class children in their own self interest*. Albany: State University of New York Press.

Flood, C., and Dorney, J. (1997). Breaking gender silences in the curriculum: A retreat intervention with middle school teachers. *Educational Action Research* 5(1): 71–86.

Ford Foundation. (1979). *Financial support of women's programs in the 1970s: A review of private and government funding in the United States and abroad*. New York: Foundation Center.

Francis, B. (2000). *Boys and girls and achievement: Addressing classroom issues*: London: RoutledgeFalmer.

Fry, C. (1990). "Sex related differences in mathematical achievement: Learning style factors." Paper presented at the annual meeting of the American Educational Research Association, Boston, MA, April.

Fuller, K. W. (2001). " 'Cool and calm inquiry': Women and the American Social Science Association, 1865–1890." Thesis (PhD) Indiana University.

Galvin, K. (1990). *Far from done: The challenge of diversifying philanthropic leadership*. New York: Women and Foundations/Corporate Philanthropy.

Gambone, M. A., and Arbreton, A. J. (1997). *Safe havens: The contributions of youth organizations to healthy adolescent development*. Philadelphia, PA: Public/Private Ventures.

Garbarino, J. (1999). *Lost Boys: Why our sons turn violent and how we can save them*. New York: Anchor Books.

Gardner, C. B. (1994). Gender, social problems, work, and everyday philanthropy among strangers. *Perspectives on Social Problems* 6: 73–95.

Garofalo, G. (1993). *Women taking power: The quest for equality*. New York: Women and Foundations/Corporate Philanthropy.

Gibbons, J. L., Hanley, B. A., and Dennis, W. D. (1997). Researching gender-role ideologies internationally and cross-culturally. *Psychology of Women Quarterly* 21(1): 151–170.

Gilder, G. (1975). *Sexual suicide*. New York: Bantam.

Gilligan, C. (1982). *In a different voice: Psychological theory and women's development*. Cambridge, MA: Harvard University Press.

———. (2003). *The birth of pleasure*. New York: Vintage Books.

Gilmore, D. (1990). *Manhood in the making*. New Haven, CT: Yale University Press.

Ginsberg, A. (2004). *Gender in urban education: Strategies for student achievement*. Portsmouth, NY: Heinemann Press, Girls Incorporated.

Gite, L. (1985). Black men and stress. *Essence* 130(November): 25–26.

Goetz, A. M. (Ed.) (1997). *Getting institutions right for women in development*. London: Zed Books.

Gordon, L. (1993). *Women's visions of welfare*. Indianapolis, IN: Indiana University Center on Philanthropy.

Grady, J., and Aubrun, A. (2000). *Talking gender equity and education: A FrameWorks message memo*. Report commissioned by the Caroline and Sigmund Schott Foundation. Cambridge, MA: FrameWorks Institute.

Greenstein, S., and Spiller, P. (1996). *Estimating the welfare effects of digital infrastructure*. Cambridge, MA: National Bureau of Economic Research.

Gunderson, L. G., Jr. (2004). "Elite young women, community, and reform: A history of the Jackson Cotillion Club, 1935–1965." Thesis (PhD) University of Memphis (Tennessee).

Gurian, M. (1996). *The wonder of boys*. New York: Jeremy P. Tarcher/Putnam.

Hall, P. (1997). Epilogue: Schooling, gender, equity and policy. In B. J. Bank and P. M. Hall (Eds.), *Gender equity and schooling*. New York: Garland Publishing.

Hargreaves, A. (1994). *Changing teachers, changing times: Teachers' work and culture in the post-modern age*. New York: Teachers College Press.

Harris, J. R. (1998). *The nurture assumption*. New York: Touchstone Books.

Hawley, R. (1991). About boys' schools: A progressive case for an ancient form. *Teachers College Record* 92(3): 433–444.

Heckler, M. (1985). *Report of the secretary's task force on black and minority health*. Bethesda, MD: U.S. Department of Health and Human Services.

Henderson, K. A. (1995). Inclusive physical activity programming for girls and women. *Parks and Recreation* 30(3): 70–78.

Hess, F. (1999). *Spinning wheels: The politics of urban school reform*. Washington, DC: Brookings Institution Press.

Heward, C. (1996). *Making a man of him: Parents and their sons' careers at an English public school 1929–1950*. London: RoutledgeFalmer.

Huff, M. (2000). " 'That benefit racket': Women and the benefit in New York theatre, 1840–1875. Thesis (PhD) University of New York (New York City).

Illinois women open up big workplace campaigns. *Responsive Philanthropy* (Fall 1991): 1.

Inter-American Development Bank. (1995). *Women in the Americas: Bridging the gender gap*. Washington, DC: Johns Hopkins University Press.

Irvine, J. M. (1994). *Sexual cultures and the construction of adolescent identities*. Philadelphia, PA: Temple University Press.

Jahan, R. (1997). Mainstreaming women and development: Four agency approaches. In K. Staudt (Ed.), *Women, international development, and politics*. Philadelphia, PA: Temple University Press.

Kabeer, N. (1994). *Reversed realities: Gender hierarchies in development thought*. London: Verso.

Kaminski, A. R. (1999). The hidden philanthropists. *Currents* 25(2): 20–25.

Katz, J., and Earp, J. (1999). *Tough guise teaching guide: Violence, media, and the crisis in masculinity* (video). Northampton, MA: Media Education Foundation.

Katz, M. B. (1989). *The undeserving poor*. New York: Pantheon Books.

Kaufman, M. (1993). *Cracking the armor*. Toronto: Penguin Books.

Keating, D. P. (1990). Adolescent thinking. In S. S. Feldman and G. R. Elliott (Eds.), *At the threshold: The developing adolescent*. Cambridge, MA: Harvard University Press.

Kimbrell, A. (1995). *The masculine mystique*. New York: Ballantine Books.

Kimmel, M. S. (Ed.) (1990). *Men confront pornography*. New York: Meridian.

———. (1996). *Manhood in America*. New York: Free Press.

Kimmel, M. S., and Messner, M. S. (Eds.) (1995). *Men's lives*, 3rd edition. Boston: Allyn & Bacon.

Kimmel, M. S. (2003). I'm not insane; I am angry: Adolescent masculinity, homophobia and violence. In M. Sadowski (Ed.), *Adolescence at school: Perspectives on youth, identity and education*. Cambridge, MA: Harvard Education Press.

Kindlon, D., and Thompson, M. (1999). *Raising Cain*. New York: Ballantine Books.

King, S. (2004). Pink ribbons inc: Breast cancer activism and the politics of philanthropy. *International Journal of Qualitative Studies in Education* 17(4): 473–492.

Kleinfeld, J. (1998). "The myth that schools shortchange girls: Social science in the service of deception." Paper prepared for The Women's Freedom Network. http://www.uaf.edu/northern/schools/myth.html (accessed May 3, 2000).

Kline, B. E. (1991). Changes in emotional resilience: Gifted adolescent females. *Roeper Review* 13(3): 118–121.

Kramer, W. M, and Stern, N. B. (1987). A woman who pioneered modern fundraising in the west. *Western States Jewish History* 19(4): 335–345.

Lakoff, George. (2002). *Moral politics: How liberals and conservatives think*. Chicago: University of Chicago Press.

Lawrence, S., Gluck, R., and Ganguly, D. (2001). *Foundation giving trends*. New York: Foundation Center.

Leadbeater, B., and Way, N. (1996). *Urban girls: Resisting stereotypes, creating identities*. New York: New York University Press.

Lee, C., and Owens, R. G. (2002). *The psychology of men's health*. Buckingham, UK: Open University Press.

Leman, C. (1977). Patterns of policy development: Social security in the United States and Canada. *Public Policy* 25: 26–291.

Liberal donors dying off, women funders more key. *Responsive Philanthropy* (Fall 1995): 5.

Liborakina, M. (1996). Women's voluntarism and philanthropy in pre-revolutionary Russia: Building a civil society. *Voluntas* 7: 397–411.

Linn, M. C., and Hyde, J. S. (1989). Gender, mathematics and science. *Educational Researcher* 18(8): 17–19, 22–27.

Lloyd, B., and Duveen, G. (1992). *Gender identities and education: The impact of starting school*. Harvester Wheatsheaf: St. Martin's Press.

Lothrop, G. R. (1989). Strength made stronger: The role of women in Southern California philanthropy. *Southern California Quarterly* 71(2/3): 143–194.

Luddy, M. (1995). *Women and philanthropy in nineteenth-century Ireland*. Cambridge, UK: Cambridge University Press.

———. (1996). Women and philanthropy in nineteenth-century Ireland. *Voluntas* 7: 350–364.

Lytle, L. J., Bakken, L., and Ronig, C. (1997). Adolescent female identity development. *Sex Roles* 37(3/4): 175–185.

Maccoby, E. (1998). *The two sexes*. Cambridge, MA: Belknap Press of Harvard University Press.

MacKinnon, A., and Allen, K. (1998). Allowed and expected to be educated and intelligent: The education of Quaker girls in nineteenth century England. *History of Education* 27(4): 391–402.

Martens, R. (1988). Youth sport in the USA. In F. L. Smoll, R. A. Magill, and M. J. Ash (Eds.), *Children in sport* (17–23). Champaign, IL: Human Kinetics.

Martinez, E. (1995). Distorting Latino history: The California textbook controversy. In D. Levine, R. Lowe, R. Peterson, and R. Tenorio (Eds.), *Rethinking schools: An agenda for change* (100–108). New York: New Press.

Mathes, V. S. (1983). Indian philanthropy in California: Annie Bidwell and the Mechoopda Indians. *Arizona and the West* 25(2): 153–166.

McCarthy, K. D. (Ed.) (1990). *Lady bountiful revisited: Women, philanthropy, and power*. New Brunswick, NJ: Rutgers University Press.

———. (1993). *Women's culture: American philanthropy and art, 1830–1930*. Chicago: University of Chicago Press.

McCarthy, K. D. (1996). Women and philanthropy. *Voluntas* 7: 331–335.

———. (Ed.) (2001). *Women, philanthropy, and civil society*. Bloomington, IN: Indiana University Press.

———. (2003). "The Ms. Foundation: A Case Study in Feminist Fundraising." Center on Philanthropy and Civil Society, The City University of New York, http://www.philanthropy.org/publications/online_publications.html.

McClelland, G. A. (2000). "Evangelical philanthropy and social control or emancipation feminism? A case study of Fisherwick Working Women's Association, 1870–1918 (Northern Ireland)." Thesis (PhD) Queen's University of Belfast (United Kingdom).

McCune, M. (2000). "Charity work as nation-building: American Jewish women and the crises in Europe and Palestine, 1914–1930." Thesis (PhD) Ohio State University.

McFate, K., Lawson, R., and Wilson, W. J. (Eds.) (1995). *Poverty, inequality and the future of social policy*. New York: Russell Sage Foundation.

McIlnay, D. P. (1998). *How foundations work*. San Francisco: Jossey-Bass.

McIntosh, P. (1988). *White privilege and male privilege: A personal account of coming to see correspondences through work in women's studies*. Wellesley, MA: Center for Research on Women.

McTighe, M. J. (1986). "True philanthropy" and the limits of the female sphere: Poor relief and labor organizations in antebellum Cleveland. *Labor History* 27(2): 227–256.

Mead, M. (1994). *Worlds apart: Missed opportunities to help women and girls*. A 1993 study of corporate and foundation giving to women's and girls' programs. Medford, MA: Lincoln Filene Center, Tufts University.

———. (2001). *Gender matters: Funding effective programs for women and girls*. Report, Women's Funding Network Month, October. Medford, MA: Tufts University.

Miedzian, M. (1991). *Boys will be boys*. New York: Doubleday.

Miller, K. (2001). "Everyday victories: The Pennsylvania State Federation of Negro Women's Clubs, Inc., 1900–1930. Paradigms of survival and empowerment." Thesis (PhD) Temple University (Philadelphia, PA).

Mindry, D. L. (1999). " 'Good women': Philanthropy, power, and the politics of femininity in contemporary South Africa." Thesis (PhD) University of California (Irvine).

Montgomery, G. (1994). Women and philanthropy, 1880–1915. *Essays in Economic and Business History* 12: 372–382.

More grantmakers are women, minorities, says Council report. *Responsive Philanthropy* (Spring 1996): 5.

Moser, Caroline O. N. (1993). *Gender planning and development*. London: RoutledgeFalmer.

Ms. Foundation for Women. (2000). *The new girls' movement: New assessment tools for youth programs*. Report and evaluation tool kit. New York: Ms. Foundation for Women.

Mulqueen, M. (1992). *On our own terms: Redefining competence and femininity*. Albany: State University of New York Press.

National Center for Education Statistics. (1990). *A profile of the American eighth grader: NELS: 88 student descriptive summary.* National Education Longitudinal Study of 1988 (NCES 90–458). Washington, DC: Office of Educational Research and Improvement, U.S. Department of Education.

———. (1991). *The state of mathematics achievement: NAEP's 1990 assessment of the nation and the trial assessment of the states.* Washington, DC: National Center for Education Statistics, Office of Educational Research and Improvement, U.S. Department of Education.

National Center for Schools and Communities. (2002). *Unlocking the schoolhouse door: The community struggle for a say in our children's education.* New York: Fordham University.

National Council for Research on Women. (1994). Do "universal" dollars reach women and girls? *Issues Quarterly* 1: 1–5.

National Science Foundation. (1990). *Assessing student learning: Science, mathematics and related technology instruction at the precollege level in formal and informal settings.* Program solicitation and guidelines. Washington, DC: National Science Foundation.

New, C. (2001). Oppressed and oppressors? The systematic mistreatment of men. *Sociology* 35(3): 729–748.

New York Women's Foundation. (1996). *The status of programming for girls aged 9–15 in New York City.*

Nicholson, H. J. (1992). "Gender issues in youth development programs." A paper commissioned by the Carnegie Council on Adolescent Development.

Noguera, P. 2005. The Racial Achievement Gap: How Can We Assure an Equity of Outcomes? In Laurie Johnson, Mary E. Fin, and Rebecca Lewis (Eds.), *Urban Education with an attitude* (11–20). Albany: State University of New York Press.

Oates, M. J. (1990). The role of lay women in American Catholic philanthropy, 1820–1920. *U.S. Catholic Historian* 9(3): 249–260.

Odendahl, T. J., and Fischer, M. (1996). *Gender and the professionalization of philanthropy.* Indianapolis, IN: Indiana University Center on Philanthropy.

Ogbu, J. (1987). *Minority education and caste: The American system in cross-cultural perspective.* New York: Academic Press.

Ogbu, J., and Simmon, H. (1998). Voluntary and involuntary minorities: A cultural-ecological theory of school performance with some implications for education. *Anthropology of Education Quarterly* 29(2).

Orenstein, P. (1994). *School girls: Young women, self-esteem and the confidence gap.* New York: Doubleday.

Osborne, J. W. (1997). Race and academic disidentification. *Journal of Educational Psychology* 89: 728–736.

Osherson, S. (1986). *Finding our fathers.* New York: Fawcett Columbine.

———. (1992). *Wrestling with love.* New York: Fawcett Columbine.

Ostrander, S. A. (2004). Moderating contradictions of feminist philanthropy: Women's community organizations and the Boston Women's Fund, 1995 to 2000. *Gender and Society* 18(1): 29–46.

Ostrower, F. (1992). *Elite insiders and outsiders: consequences for philanthropy.* New Haven, CT: Yale University Program on Non-Profit Organizations.

———. (1995). *Why the wealthy give: The culture of elite philanthropy.* Princeton, NJ: Princeton University Press.

Pastor, J., McCormick, J., and Fine, M. (1996). "Makin' " homes: An urban girl thing. In B. J. Ross Leadbeater and N. Way (Eds.), *Urban girls: Resisting stereotypes, creating identities* (15–34). New York: New York University Press.

Pittman, K. (1991). *Bridging the gap: A rationale for enhancing the role of community organizations promoting youth development.* Report. Center for Youth Development & Policy Research, Academy for Educational Development.

Piven, F. P., and Cloward, R. A. (1982). *The new class war.* New York: Pantheon Books.

Pleck, J. H. (1981). *The myth of masculinity.* Cambridge, MA: MIT Press.

Pleck, J. H., Lund Sonenstein, F., and Ku, L. C. (1993). Masculinity ideology and its correlates. In S. Oskamp and M. Costanzo (Eds.), *Gender issues in contemporary society* (85–110). Newbury Park: Sage Publications.

Plemper, E. (1996). Women's strategies in Dutch philanthropy. *Voluntas* 7: 365–382.

Pollack, W. S. (1998). *Real boys: Rescuing our sons from the myths of boyhood.* New York: Random House.

Powell, K. (1992). The sexist in me. *Essence Magazine*, September.

———. (2000). Confessions of a recovering misogynist. *Ms. Magazine*, January.

Preston, M. (1999). " 'The unobtrusive classes of the meritorious poor': Gentlewomen, social control and the language of charity in nineteenth-century Dublin." Thesis (PhD) Boston College.

———. (2004). *Charitable words: Women, philanthropy and the language of charity in nineteenth-century Dublin.* Praeger Publishers.

Prochaska, F. K. (1980). *Women and philanthropy in nineteenth-century England.* Oxford: Clarendon Press.

Rao, A., Anderson, M. B., and Overholt, C. A. (Eds.) (1991). *Gender analysis in development planning.* West Hartford, CT: Kumarian Press.

Raughter, R. (1997). A discreet benevolence: Female philanthropy and the Catholic resurgence in eighteenth-century Ireland. *Womens History Review* 6(4): 465–484.

Reichert, M. (2001). Rethinking masculinities: New ideas for schooling boys. In W. Martino and B. Meyenn (Eds.), *What about the boys?* (38–52). Philadelphia, PA: Open University Press.

Research for Action. (1996). *Girls in the middle: Working to succeed in school.* Washington, DC: American Association of University Women Educational Foundation.

Richards, S. L. (1996). Library philanthropy with a personal touch: Phoebe Apperson Hearst and the libraries of Lead and Anaconda. *Libraries & Culture* 31(1): 197–208.

Riordan, C., and Lloyd, S. (1990). Resolved: Many students, especially women, are best served by single-sex schools and colleges. *Debates on Education Issues* 2(3): 1–7.

Roach, C., Yu, H. C., and Lewis-Carp, H. (2001). Race, poverty and youth development. *Poverty & Race* 10(4).

Ross Leadbeater, B. J., and Way, N. (1996). *Urban girls: Resisting stereotypes, creating identities*. New York: New York University Press.

Rothenberg, D. (1995). Supporting girls in early adolescence. *Eric Digest*, October.

Sabo, D., and Gordon, D. F. (Eds.) (1995). *Men's health and illness*. Thousand Oaks, CA: Sage Publications.

Sadker, M., and Sadker, D. (1994). *Failing at fairness: How America's schools cheat girls*. New York: Charles Scribner's Sons.

Salamone, R. C. (2003). *Same, different, equal*. New Haven, CT: Yale University Press.

Sarnoff, J. D. (2003). "White women and respectability in antebellum New Orleans, 1830–1861 (Louisiana)." Thesis (PhD) University of Southern Mississippi.

Sandage, S. A. (1999). Gender and the economics of the sentimental market in nineteenth-century America. *Social Politics* 6(2): 105–130.

Saveth, E. N. (1980). Patrician philanthropy in America: The late 19th and early 20th centuries. *Social Service Review* 54(1): 76–91.

Savin-Williams, R. (2005). *The "new" gay teen: Post-gay and gayishness among contemporary teenagers*. Cambridge, MA: Harvard University Press.

Schorr, L. B. (1988). *Within our reach: Breaking the cycle of disadvantage*. New York: Anchor Books.

Schur, E. M. (1984). *Labeling women deviant: Gender, stigma, and social control*. New York: Random House.

Servatius, M. (1992). *Shortsighted: How Chicago-area grantmakers can apply a gender lens to see the connections between social problems and women's needs*. Chicago: Chicago Women in Philanthropy.

Sewell, T. 1998. Loose canons: Exploding myths of the "black macho" lad. In Debbie Epstein, Jannette Elwood, Valerie Hay, and Janet Maw (Eds.), *Failing boys? Issues in gender and achievement*, (111–127). Philadelphia, PA: Open University Press.

Shapiro, J. P., Parssinen, C., and Brown, S. (1992). Teacher-Scholars: An action research study of a collaborative feminist scholarship colloquium between schools and universities. *Teacher and Teacher Education* 8(1): 91–104.

Shapiro, J. P., Sewell, T. E., and DuCette, J. P. (1995). *Reframing diversity in education*. Lancaster: Technomic Publishing Company.

Shaw, S., and Taylor, M. (1995). *Reinventing fundraising: Realizing the potential of women's philanthropy*. San Francisco: Jossey-Bass.

Shelton, C. L. (2003). "We are what we do: The national program of Alpha Kappa Alpha Sorority, Incorporated. A post-modern corporatist interpretation of African-American women's philanthropy." Thesis (PhD) University of Kentucky.

Silverstein, O. (1994). *The courage to raise good men*. New York: Viking.

Skelton, C. (2001). Typical boys?: Theorizing masculinity educational settings. In B. Francis and C. Skelton (Eds.), *Investigating gender: Contemporary perspectives in education* (164–176). Philadelphia, PA: Open University Press.

Skocpol, T. (1991). Targeting within universalism. In C. Jencks and P. Peterson (Eds.), *The urban underclass* (411–435). Washington, DC: Brookings Institution Press.

Sommers, C. H. (2000). *The war against boys: How misguided feminism is harming our young men*. New York: Simon & Schuster.

Spencer, M. B., and Dornbusch, S. M. (1990). Challenges in studying minority youth. In S. S. Feldman and G. R. Elliott (Eds.), *At the threshold: The developing adolescent* (123–146). Cambridge, MA: Harvard University Press.

Staudt, K. (Ed.) (1997). *Women, international development, and politics*. Philadelphia, PA: Temple University Press.

Stillion, J. M. (1995). Premature death among males. In D. Sabo and D. F. Gordon (Eds.), *Men's health and illness*. Thousand Oaks, CA: Sage Publications.

Stoltenberg, J. (1990). *Refusing to be a man*. London: Fontana.

Stoudt, B. G. (2006). "You're either in or you're out": School violence, peer discipline and the (re)production of hegemonic masculinity. *Men and Masculinity* 8(3): 273–287.

Summers, L. (1992). The most influential investment. *Scientific American* 132.

Sundar, P. (1996). Women and philanthropy in India. *Voluntas* 7: 412–427.

Swain, S. (1996). Women and philanthropy in colonial and post-colonial Australia. *Voluntas* 7: 428–443.

Tannen, D. (1990). *You just don't understand*. New York: Ballantine Books.

Taylor, L. C., Jr. (1963). Josephine Shaw Lowell and American philanthropy. *New York History* 44(4): 336–364.

Taylor, J. M., Gilligan, C., and Sullivan, A. M. (1995). *Between voice and silence: Women and girls, race and relationship*. Cambridge, MA: Harvard University Press.

Thomas, D. (1993). *Not guilty: The case in defense of men*. New York: William Morrow and Company.

Thorne, B. (1993). *Gender play: Girls and boys in school*. New Brunswick, NJ: Rutgers University Press.

Thorne-Murphy, L. (2001). " 'Art's a service': Women's philanthropy and the role of the author in mid-Victorian England (Charles Dickens, Charlotte Yonge, Charlotte Tucker, Elizabeth Barrett Browning)." Thesis (PhD) Brandeis University.

Three Guineas Fund. (2001). *Improving philanthropy for girls' programs*. Report. San Francisco, CA: Three Guineas Fund.

Twenge, J. M. (1997). Changes in masculine and feminine traits over time: a meta-analysis. *Sex Roles* 36(5/6): 305–325.

Tyre, P. (2006). The trouble with boys. *Newsweek*, January 30, 44–52.

U.S. Preventive Services Task Force. (1996). *Guide to clinical preventive services*, 2nd edition. Baltimore, MD: Williams & Wilkins.

Valentine Foundation. (1990). *A conversation about girls*. Conference Proceedings. Philadelphia, PA: Valentine Foundation.

Viladrich, A., and Thompson, A. (1996). Women and philanthropy in Argentina: from the society of beneficence to Eva Peron. *Voluntas* 7: 336–349.

Von Schlegell, A. (Ed.) (1993). *Women as donors, women as philanthropists.* San Francisco: Jossey-Bass.
Waldron, I. (1976). Why do women live longer than men? *Journal of Human Stress* 2: 1–13.
Walkerdine, V. (1990). *Schoolgirl fictions.* New York: Verso.
Walton, A. (2005). *Women and philanthropy in education.* Bloomington, IN: Indiana University Press.
Ward, J. V., Rothenberg, M., Benjamin, B. C., and Feigenberg, L. (2002). *Results from focus groups and interviews with gender equity consultants.* Report commissioned by the Caroline and Sigmund Schott Foundation, Cambridge, MA: Caroline and Sigmund Schott Foundation.
Way, N., and Chu, J. (Eds.) (2004). *Adolescent boys: Exploring diverse cultures of boyhood.* New York: New York University Press.
Weber, L. (1998). A conceptual framework for understanding race, class, gender, and sexuality. *Psychology of Women Quarterly* 22: 13–32.
Williams Elliott, D. (2002). *The angel out of the house: Philanthropy and gender in nineteenth-century England.* Charlottesville, VA: University of Virginia Press.
Wilson, W. J. (1997). *When work disappears: The work of the new urban poor.* New York: Alfred A. Knopf.
Winter, M. F., and Robert, E. F. (1980). Male dominance, late capitalism and the growth of instrumental reason. *Berkeley Journal of Sociology* 249–280.
Women and philanthropy analyzed; national action agenda developed. *Responsive Philanthropy* (Winter 1994): 6.
Women Working in Philanthropy. (1990). *Doubled in a decade, yet still far from done: A report on awards targeted to women and girls by grantmakers in the Delaware Valley.* Philadelphia, PA: Delaware Valley Grantmakers.
Womens Way USA: Expanding successful Philadelphia fund raising to women's nonprofits nationwide. *Responsive Philanthropy* (Winter 1990): 4.
Zeldin, S., Kusgen McDaniel, A., Topitzes, D., and Clavert, M. (2000). "Youth in Decision-Making: A Study on the Impacts of Youth and Adults in Organizations." Commissioned by the Innovation Center for Community and Youth Development, National 4-H.

Index

A Conversation About Girls, 1, 90, 229, 230, 233, 253
A Nation at Risk, xii
accountability, xiii, 7, 89, 112, 150, 204, 207
   district level accountability, 113
   systems, 113, 116, 220
   language of, 147
   stakeholder groups, 208
   creating it, 173
   relationship with evaluation, 208, 210, 215, 217
   and diversity, 208
   and "No Child Left Behind," 113
   and uniformity, 208
   process of change, 210
   and funding, 211, 215, 217
   who to blame, 220
ADD, 136
ADHD, 118
African Americans, ix
   boys, 6, 11, 109–116, 126, 140, 141, 143, 152, 187, 228
   early education of, ix
   girls, 94, 95, 114
   hair, 196, 201
Agency Diversity Data Form, 82
AIDS, 47, 50, 70, 78, 87
Algebra Project, The, 114
Alice Paul Leadership Center, 207
Alliance on Gender, Culture, and School Practice, 140, 216
Amaro, Hortensia, 91, 103
American Association of University Professors (AAUP), 2
American Association of University Women (AAUW), 2, 61
Annenberg Foundation (Walter), 217, 236

Annie E. Casey Foundation, 8
Appalachian Women's Leadership Project, 163
Asian Immigrant Women Advocates, 172
assessment, 13, 208, 227
   needs assessment, 7, 100
   gender focused, 13, 227
   accountability, 13
   needs assessment, 100
   shared assessment, 171
   high-stakes testing, 207
   and inquiry, 226
   program assessment, 8, 11
   variety, 207
   tools, 224
   feminist assessment, 226

Barnard College, ix
Blocks Together, 168
body image, 193
boys
   war against, x, 4
   defective, 18, 22, 25
   more attention to, 2, 21, 41
   boys and their mothers, 4
   treating boys like girls, 5, 46
   treating boys differently than girls, 18, 22, 25, 42, 49, 53, 55–57, 69, 86, 91, 124, 137, 206, 214
   diversity of, 4, 11, 126
   definitions of masculinity, 4
   and violence, 4
   African American boys, 6, 11, 110–116, 140, 210, 228
   "boys will be boys" attitude, 11, 121, 188, 221
   Boys' self-esteem, 11, 19

boys—*continued*
  Boys are "defective," model, 18, 22, 25–26, 31
  and paying attention in school, 21, 24, 209
  funding of, 40
  math and science, 48
  and violence, 49, 58, 124–125
  separate programs for boys and girls, 54
  boys' clubs, 54–55
  high school dropouts, 116
  and "achievement gap," 118
  "invisible curriculum," 120
  peer relationships, 127
  mentoring boys, 128
  resilience, 133–134
  funding for, 138
  homophobia, 139
Boys and Girls Center, 52, 53
Brown, Lyn, 102
Bush, Laura, 4

Capek, Mary Ellen, 33, 74
Center for the Study of Boys' Lives, 117, 122
Center for Young Women's Development, 169
Chicago Women in Philanthropy, 81
Chicago Youth United, 169
Children and the Media Program, 144
Children Now, 144
City Action Teens Teams (CATT), 163
civic engagement, xi
Clarke, Edward, ix
Clearsighted protocol, 81
Cohn, Marjorie, 145
Collaborative Fund for Healthy Girls/Healthy Women, 160, 175, 222, 230
Collaborative Fund for Youth-led Change (CFYS), 157, 158–174, 239
Colorado Progressive Coalition, 173
Columbia University, ix
Columbine High School, 147
Conservative, 31, 32, 121, 136
Cook, Donelda A., 90
Cultural Logic, 17
Cultural Models Research, 17
Curriculum, 2, 22, 76, 109, 113, 117, 168, 179, 207–208, 211, 216
  Gender within, 2, 6
  Engaging students in, 3
  GATE program, 3
  "add women and stir," approach, 2, 5, 209
  Replication, 216
  Sexist curriculum, 107
  And community, 179
  "Young Women for Change," 180
  For social justice, 195
  Teacher's role in developing it, 3
  And women, 5, 180, 209
  Hidden curriculum, 107, 152
  For boys, 119, 120, 123
  And diversity, 195, 197, 208
  And accountability, 204
  High stakes testing, 204
  And assessment, 208

Diversity, 3, 22, 43, 76–87, 93, 158, 160, 164–167, 170–171, 180–182, 208, 227, 233, 236, 242, 244, 251
  SEED program, 3
  cultural diversity, 22, 93
  racial diversity, 43
  and grantmaking, 75–87
  "Agency Diversity Data Form," 82
  in the workforce, 82
  of youth identities, 160
  Accountability and, 209
Dodge, Grace, xi
domestic violence, 1, 50, 98, 148

Education Development Center, 190
Education Trust, The, 113

Equal Voices, 56, 57
evaluation, 7, 10–11, 13–14, 42, 46, 84–85, 100, 116, 167, 170, 190, 194, 200, 203–231
 quantitative, 7
 qualitative, 7, 205, 215, 221
 earmarking funds for, 7, 204
 youth participation in, 11–13, 170, 226
 opportunities vs. causes for blame, 13
 participatory evaluation, 14, 194–195, 221–222
 ethnography, 221
 analytic lenses, 85
 mentoring programs, 100
 evaluation teams, 190
 diversity within, 190
 and urban education, 204
 educational evaluation, 204
 measure of "success" and "failure," 205, 209, 226
 process and product evaluation, 207
 and "No Child Left Behind," 207
 and accountability, 208
 impact on students, 218
 VACO, 223
 ISM, 224
 GATE program, 220
 triangulation of data, 226
 external evaluation, 229
 and flexibility of, 231

Feminism, 139, 184, 192
 Feminist assessment, 226
Fine, Michelle, 91
Ford Foundation, 10, 38, 96, 159, 236
The Foundation Center, 38, 61
Foundation News, 38
Frameworks Institute, 212
Funders Collaborative on Youth Organizing, 159
Funding, 7, 8, 9, 11–14, 33–45, 90–98, 103, 138–139, 175, 179, 180–181, 187, 190–238

 of education initiatives, vii
 priorities, vii
 strategic funding, viii
 women's issues, 1, 103
 youth voice in, 7–8
 diversity within, 9
 mutli-year funding, 14
 agencies, 1, 13
 gender category, 2, 36–37
 Women's Funding Network, 39
 equality, 6, 96, 116
 process, 7
 collaboration, 8, 10, 92
 evaluation, 12, 203–238

gang violence, 58
gender analysis, 10, 34–37, 47, 68–76, 81–86, 123
Gender Awareness Through Education (GATE), 3, 188–189, 217–220, 236
gender equity, 5, 8–9, 13–14, 17–18, 21–30, 41, 44, 76–77, 79, 83, 87, 180–187, 203–231
 and boys, 4
 early supporters of, ix
 in education, xiii
 philanthropic support of, xii, xiii
 definitions of, xv, 20, 189
gender lens, 9, 35, 61, 67–87
gender norms, 35, 52, 57, 160, 214
gender-specific programs, 36–37, 40, 46, 50
Gilligan, Carol, 102, 134–136, 188
Girls
 educational equity, 2, 5, 6, 70
 grantmaking, vii, 1, 2, 10, 33–46, 72, 90, 92, 96–98, 105–106, 177, 185
 as grantmakers, 7–8, 12, 13, 177–180, 185
 "disadvantaged" model, 18, 21, 22, 25–26, 31, 54, 67, 177, 213, 228
 affirmative action, 21
 in math and science, 22

Girls—*continued*
    sports, 53
    computers, 54
    discrimination against, 19, 187
    favoring girls, 19, 213
    self-esteem/leadership, 91, 102, 177, 180, 185, 192–193, 209, 222–223
    relationships to race and class, 10, 13, 60, 92–99, 101–105, 212, 213, 228
    program evaluation, 13, 14, 217, 222–226
    programs for, 12, 14, 36–37, 40, 46–56, 74–75, 89, 100, 105, 162–173, 222
    safety/sexual harassment, 49, 99, 101, 105, 160, 196, 201
    co-education, 51
    public policy, 50
    media, 145
    female friendships, 99
    girls' schools, 117
    health care, 70
    "at risk," 84, 205
    gender norms, 57
    listening to girls' voices, 10, 90, 102, 160, 223
    relationship with/to boys, 4, 9, 11, 14, 18, 21–30, 34, 47–48, 52–56, 69, 75–76, 86, 91, 114, 119–133, 187, 206, 214, 221
    math and science, 48, 117, 188, 206, 212
    early education, viii, ix, x
    diversity among them, x, xiii, 12, 180
Girls' Action Initiative, The, 207
Girl's Best Friend Foundation, 12, 191–201, 235
Girls Coalition of Greater Boston, 14
Girls Incorporated (Inc.), 51, 55, 207
Girls Resiliency Program (GRP), 163
Girl Scouts of America, 207
Girl Scouts of Milwaukee, 163

Girl World, 12, 191–195
Global Business Network, 113
Grant Craft, 10
Greater Boston Study, 40–43

Hair Project, The, 196
Harvard Civil Rights Project, 113
Harvard University, ix, 91, 140, 142, 216, 238
Haviland, Tessa, 184
Hip-hop, 143
HIV, 67
Homosexuality, 147
    Bullying, 53, 187, 189
Howie, Sarah, 182
human rights, 67, 150, 235, 239
Hyams Foundation, 82, 237

Immigrants, 74
Ingram, Helen, 49
International Storytelling Measure (ISM), 22, 224–226
Irvine Foundation (James), 8

Kaplan, Elaine, 91, 104
Katz, Jackson, 144, 147–148
Kellogg Foundation, 8
Khmer Girls in Action (KGA), 162–163
Kimmel, Michael, 121, 136–139, 147

Lakoff, George, 23, 27
Land Equity Campaign, 164
Latina/s, 43, 94, 98–99, 157, 163
Latino/s, 100, 114, 228, 233
Liberal, 31–32
Lyons, Mary, viii

Males are the Model, 56–59
Malloy, Emily, 182
Mann, Horace, vii
Maracek, Jeanne, 90, 101
Mark, Shirley, 8, 12, 187–190
masculinity, 4, 11, 65, 118, 121–154, 207, 228
    Monolithic masculinity, 118

Massachusetts Coalition for Occupational Safety and Health, 169
materialism, 140
Mayors Commission on Women in Philadelphia, 90
McAdoo, Harriette Pipes, 90, 93
McIntosh, Peggy, 91, 105, 107, 188, 190
McPherson, Donald, 148
Mead, Molly, 33, 73–74
media, 4, 11, 20, 25, 100, 106, 110, 118, 125, 133–134, 141–145, 152–153, 187, 193–196, 236
melting pot, 32
Men Stopping Violence, 149–150
Mentoring, 11, 12, 95, 100, 149, 157, 159, 171–172, 209
Mentors in Violence Prevention Program (MVP), 148–150
Michigan Women's Foundation, 12, 177–186, 230
Miller, Patti, 144
Moore, Melody, 182
Moses, Robert, 114
Mount Holyoke, viii
Ms. Foundation, 1, 7, 11–12, 133–134, 157, 160–164, 170, 174, 205, 216–217, 221–226

National Center for Schools and Communities, 204
National Council of Research on Women, 9
National Football League (NFL), 148
National Organization for Men Against Sexism, 136
National Writing Project, 3
Native Americans, 97
Neighaver, Bethany, 184
Neighborhood House, 53
New York City, 55, 210
New York Women's Foundation, 55
Nickelodeon, 145

No Child Left Behind (NCLB) Act, xii, xiii, 3, 7, 111, 113, 204, 207, 236
Media scrutiny, vii
Noguera, Pedro, 151–152
Nuriddin, Sulaiman, 149–150

Palmert, Jessica, 194
Peace Corps, 67, 183
Pearls for Teen Girls, 169
Pennsylvania Humanities Council, 188, 217
Phi Delta Kappa, 113
Philadelphia Public Schools, 188
Pleck, Joseph, 137
Pollack, William S., 142
Poverty, 44, 47, 81, 94, 98, 152, 164, 188, 213, 215, 228, 244
Powell, Colin, 110
Powell, Kevin, 143
professional development, 3, 14, 188, 189–190, 207, 210, 227–228, 236
 funding for, 14
 best practices, 188
 importance of, 189
 and self-reflection, 207
 GATE program, 236

race, 5, 9–11, 14, 19, 29–30, 35, 43, 45, 60–61, 68–70, 74, 82–84, 85–91, 97–98, 101, 109, 112, 115, 118, 138, 141, 152, 153, 157, 160–171, 175, 180, 188, 190, 209, 219, 221, 228, 229, 231, 238
Radcliffe College, ix
Ransom, Jane, 90, 96
Resilience, 11, 133, 151, 247
Respect Me, Don't Media Me!, 193
Ritalin, 23

Savin-Williams, Ritch, 146–147
SAT, 118
Schneider, Anne, 49, 50

Schott Foundation (Caroline and Sigmund), 17, 21, 26, 31, 32, 112, 114, 116, 187–189, 212, 216, 236–238
Schwarzenegger, Arnold, 147
Science, 2, 22, 45, 112, 118, 134, 142, 146, 188, 206–207, 212
  women and, vii, xiii
Seeking Educational Equity and Diversity (SEED), 3
Sexuality, 5, 9, 35, 60–61, 66, 101, 103, 139, 143, 146–147, 169, 180, 192–193, 236
  premature behavior, 100
  sexual abuse, 98, 102, 105
  date rape, 102, 147
  pregnancy, 97, 103–105
Sista II Sista, 172
Sisters Empowering Sisters, 12, 191–202
Sister in Action for Power, 163–164
Smith, Rosa, 11
Smith, Sophia, ix
social justice, 9, 11, 13, 69, 169, 173, 195, 197, 205, 222, 230
Sommers, Christina, 136
special education, 111, 113–114, 219
spectrum of social change, 161–162
spirituality, 90, 100
"strict father" model, 23, 27
Stallone, Sylvester, 147
Summer, Lawrence, viii
Surdna Foundation, 159
Sweeney, Margaret, 182

Take Our Daughters to Work, 161
Take Our Daughters and Sons to Work, 161
Tatum, Beverly Daniel, 188
Taylor, Jill McLean, 91, 102
Teachers
  attention to boys, 2, 21, 41
  curriculum development, 3
  and gender, 3, 217–220

professional development, 3, 189–207
gender blind, 5
gender bias, 5, 19
  favoring girls, 24–25, 213
  pre-service, 6
  standardized tests, 209
  and boys of color, 210
Teens Lead @ Work, 169–170
Television, 144–145
Three Guineas Fund, 203, 214–215
Tides Foundation, 159
Title I, 109, 115
Title IX, 6, 206, 213
Tolman, Deb, 102
Tough Guise Teaching Guide, 144, 153
Tracy, Carol E., 98
Tully, Mary Jane, 38

United Way, 183, 237

Valentine Foundation, 1, 10, 89–107, 229–230, 238, 252
Violence
  by/against men and boys, 5, 58, 119–125, 134, 142, 147–148, 169
  against women of color, 172
  African American boys, 143
  in media, 144–145
  dating/interpersonal, 173, 180
  girl on girl, 196–201
  in schools, 212
  domestic, 50, 98
  youth, 58
  prevention, 148–151, 235
  MVP program, 148–149
  and girls' self-esteem, 91
  against women, 98–99, 147–149
Voice, Action, Comportment, and Opportunity (VACO), 222–226

Ward, Janie, 103, 140
We Are All the Same, 56–59

Wefald, Susan, 134
White Ribbon Campaign, 150
Women and Foundations/Corporate Philanthropy (WAF/CP), 90, 96
Women and Philanthropy, 1, 82
Women's College Coalition (WCC), vii, viii, xiii
Women's Colleges, viii, ix, xiv
Women's Funding Network, The, 39
Women's philanthropy movement, xii
Women's Way, 89–90

Young Women's Action Team (YWAT), 222
Young Women for Change, 12, 177–185, 237
Young Women's Project, 173

GPSR Compliance

The European Union's (EU) General Product Safety Regulation (GPSR) is a set of rules that requires consumer products to be safe and our obligations to ensure this.

If you have any concerns about our products, you can contact us on

ProductSafety@springernature.com

In case Publisher is established outside the EU, the EU authorized representative is:

Springer Nature Customer Service Center GmbH
Europaplatz 3
69115 Heidelberg, Germany

www.ingramcontent.com/pod-product-compliance
Lightning Source LLC
LaVergne TN
LVHW011808060526
838200LV00053B/3703